MARTIAN SUMMER

PRAISE FOR *MARTIAN SUMMER*

"Readers will thrill to this slightly offbeat firsthand account of scientific determination and stubborn intellect. This behind-the-scenes look delivers a fascinating journey of discovery peppered with humor."

—*Publishers Weekly*

"A candid and precise account of the ups and downs of a space mission. This book shows what it is to participate in a short and intense landed Mars expedition. It gives the feel of the pressure and excitement at mission control, where engineers, managers and scientists work together while trying to satisfy contradictory requirements, showing the human side of science with refreshing honesty."

—Nilton Renno, Professor of
Atmospheric and Space sciences,
University of Michigan

"It is as if I imagined Holden Caulfield writing about the mission. *Martian Summer* is a riot."

—Peter Smith, Professor, Lunar and Planetory
Laboratory, University of Arizona, and
Principal Investigator of the Phoenix Project

"*Martian Summer* gives a picture not so much of the science of what was discovered as of the sociology of how it all happened. An informative and charming semi-insider account of how such a mission operates, how humans fare on Mars time, and how scientists and administrators behave under extreme stress. You might even find yourself, as I did, unexpectedly choking up when the mission's robot goes to sleep for the very last time."

—*The Washington Post*

"At first glance, it looks like a charming independent bookstore, a West Village gem with a window display featuring artful stacks of gleaming hardcovers. But, wait a minute. Is that one book? Like, many, many copies of the same book? The book is Mr. Kessler's account of NASA's 2008 Phoenix Mars Lander mission, reported during 90 days inside mission control, in Tucson, alongside 130 leading scientists and engineers."

— *The New York Times*

"In the summer of 2008, when the Phoenix Mars Lander beamed home news of water ice and other discoveries, Andrew Kessler was one of the first to hear about it. That's because Kessler gained access to the Phoenix mission control room, an experience that led to his new book, *Martian Summer*. We loved reading about what it was like to be so close to the action."

—*Popular Mechanics*

"An amazing mission. Kessler works to remind us of the magic of spaceflight and exploration in a manner we can all understand. What you will find is a peek behind the curtain at what makes a mission to Mars work – in all of its quirky glory."

—*American Space*

"An inside look at a Mars mission. The author provides some fascinating glimpses of the real work of a space mission: planning activities for the lander, dealing with peremptory orders from NASA, interpreting the sometimes ambiguous data and occasionally letting one's hair down for a party. Die-hard space fans will find much to keep the pages turning."

—*Kirkus*

"In the summer of 2008, Kessler lived a space dream. His offbeat and often humous book captures the real-life drama of the mission and its passionate crew as they attempt to dig up permafrost on the North Pole of Mars. This three-month epic changed our understanding of Mars and sheds new light on the importance of curiosity-based research and NASA missions."

—BoingBoing.com

"Kessler hit the astronomy jackpot when he was allowed in the control room of the Phoenix lander, the Mars probe that struck water ice. Geeky giddiness overcomes him at times, but Kessler's access allows unique insights into characters like mission leader Peter Smith with his "cowboy swagger," as well as into the challenges of talking to a robot across hours-long communication delays."

—*Discover* Magazine

"A fascinating, quirky, and humorous look inside the alternate universe of mission control, where everybody is sleep deprived from working on Martian time and employees carry bottles of their urine in Trader Joe's shopping bags. Kessler, a keen observer of social and political dynamics, reveals the conflicting mindsets and priorities of engineers, scientists, and NASA management; scientific rivalries; and even the genesis of a space conspiracy theory."

—*Library Journal*

MARTIAN SUMMER

ROBOT ARMS, COWBOY SPACEMEN, AND MY 90 DAYS WITH THE PHOENIX MARS MISSION

ANDREW KESSLER

PEGASUS BOOKS
NEW YORK

For Lottie and Bobo

MARTIAN SUMMER

Pegasus Books LLC
80 Broad St., 5th Floor
New York, NY 10004

Copyright © 2011 by Andrew Kessler

First Pegasus Books cloth edition April 2011

Images that appear on chapter and part titles as well as on pages ii-iii, and ix are
used with courtesy of NASA/JPL/UA.

Interior design by Maria Fernandez

Library of Congress Cataloging-in-Publication Data is available.

ISBN: 978-1-60598-346-2

10 9 8 7 6 5 4 3 2 1

Printed in the United States of America
Distributed by W. W. Norton & Company, Inc.

CONTENTS

AUTHOR'S NOTE

THIS IS A BOOK ABOUT MARS AND THE HUMANS THAT MAKE ROCKET science possible. This is *not* the most accurate account of this NASA mission. For that, you'll want to review the science papers or other Mars-related literature. Instead, this is an account of winning the nerd lottery: The luckiest fanboy in fandom gets a shot to spend three months with unfettered access to Mission Control. It's just your average summer trying to capture the story of 130 of the world's best planetary scientists and engineers exploring the north pole of Mars. It's a warts-and-all look at the Phoenix Mars mission from a regular guy who loves space.

ACKNOWLEDGMENTS

I T'S HEARTBREAKING NOT TO LIST EVERY MEMBER OF THE PHOENIX TEAM in these pages. They are awesome and I suggest you find them and bring them a glass of champagne to celebrate their commitment to awesomeness. I'm honored to have spent time with them.

With one giant leap of faith, Peter Smith and Catherine Patterson made this project possible. The support of the entire Phoenix team and my family made this opportunity worthwhile. Special thanks to my granny, Bella and my brother, Josh for having to hear the Mars story more than anyone. Ben Kaplan and Mike Spiegel for nearly unlimited pep talks and advice. All the people in my life who read and read and encouraged and wouldn't let me give up. Michael Wolfson, Joe Lazar, Karen Ingram, Courtney Rein, Matt Wilkens, Allison Lucas, Talia Avisar, Kym De La Roche, Andrea Thompson, Beth Mulhern, Ken Hamm, Judi Powers, Clayton Carpinter, Bernadine Lim, Kristina Grish, Frank Schaap, all the Klaristenfelds, Rasmussens and Becks. Veronica Kavass for being a sounding board, critic and booster when *Martian Summer* needed it most. Jill Swenson for her keen journalistic eye and thoughtful comments.

There would be no books without agents and publishers. Thanks to Special Agent Adam Korn, who kept me honest and argued with me about what's funny. Claiborne Hancock and Jessica Case, publisher and editor extraordinaire, not only put out this book with a lot of love, they listened and nodded to lots of Mars rants.

And thanks to the super creative folks at HUGE for making a crazy idea about an ad campaign for a Mars books a reality.

PART I

THE PHOENIX
OF TUCSON

DATE: JUNE 04, 2007

THE STORY BEGINS TWO MONTHS BEFORE THE LAUNCH OF THE Phoenix Mars Lander. One year before the landing. It takes ten months to fly at 74,000 mph to arrive on Mars. It's far.

The subject of the story is a Martian photographer.

"Don't call me that," Peter Smith, the world's greatest Martian Photographer says dryly. "It really diminishes the science."

This is a story about the world's greatest Mars picture-taker and his robot, Phoenix.

"And don't make me look like some wacko mad scientist," Peter says. He has a hard enough time with the mission's image as it is. Peter is particular about the mission's image because he knows how getting it right has the potential to inspire children and adults alike. More than half his team is here because they grew up watching Apollo and Viking missions.

"What's going to inspire the next generation?" he wants to know.

We're sitting in the back yard of Peter's Tucson home. We're getting off on the wrong foot and I can't stop imagining Peter working in his

Martian photo studio posing little aliens on the Red Planet. Stupid, I know.

Peter is intimidating. He is tall—very tall—with a shock of white hair, bushy eyebrows, big mustache, a robust Buddha-like belly and an alpha-male cowboy swagger. He towers over me and says little. Only grimacing and asking if I'm sure I'm up for the task, correcting me when I say things like "Martian photographer" or make other interplanetary gaffes. I blabber to fill the silence. It's not uncommon to feel this way when you first meet the brilliant, geeky—

"Please don't make us look like geeks, either" says the brilliant John Wayne of space.

"Go collect some firewood for dinner," he says. I do it. When I return, Peter breaks the wood with his hands, starts a small fire, and tells me a story.

JUST FIVE HUNDRED YEARS AGO, MARS WAS A DOT, A SPECK OF LIGHT. Then came the first telescopes, and Mars ceased to be a dot. It became, instead, a world which scientists claimed was much like our own. Imaginations ran wild, and before long, rather than see vast, wonderful possibilities, we feared a Martian attack. As a war of the worlds loomed, Mars became a source of fear and anxiety.

It wasn't until our first stumblings into the solar system in the 1960s, when Mariner 4 snapped photos of Mars's surface that we caught a glimpse of what it might *actually* be like. Rather than an advanced civilization poised for an attack, Mariner 4 showed us a lifeless, desolate place. A few years later, in the 1970s, Viking I confirmed those first impressions: Mars was nothing to fear. Just a dead planet; barely worth exploring. The missions stopped. The scientific dreamers lost sleep and became depressed.

Then a discovery in the 1990s changed everything. ALH84001, a piece of Mars ejected by a cosmic collision, was thrust through the solar system and somehow landed on Earth. It was found in Antarctica in 1984 but no one took much interest. When scientists at NASA finally cut it open to take a closer look, they found something shocking: evidence of life. Tiny microbes, simple little guys with evidence of a few of the basic structures of life, like a cell wall. It was the basic innards of something you might find in the extreme environments of the Earth—sulfur vents at the bottom of the ocean, the dry valleys of Antarctica, or the Andean desert. Clearly there was more to discover on Mars. So, we headed back.

PETER SMITH IS A MASTER AT CONJURING THESE LITTLE MARS VIGNETTES. That's not his only virtue or why we're here. Peter built an excavator to operate on Mars. It took five years of construction and nearly a lifetime of dreams. In a few months, he will watch a Delta II rocket blast off into space carrying his 800-pound lander with a long arm that can dig into the surface of Mars. Past Mars missions toted along soup-spoon style digging equipment, but Phoenix brings a mini backhoe to do real interplanetary digging. His mission is called The Phoenix Mars Lander. Phoenix for short.

Peter builds cameras for space. Capturing the universe on film is a great gig. He built almost half of the cameras that have operated on Mars, and got to where he is by working his way up from research assistant to Mission Captain—or Principal Investigator to NASA insiders. It's a classic photon-to-Charged-Coupled-Device story.

You might remember waking up one summer morning in 1997 to a well-cropped ocher-colored Martian landscape on the front page of your newspaper or computer screen. Remember? Peter took that image. His camera, fixed to a robot called Pathfinder, captured the alien landscape using a simple yet brilliant trick to get non-scientists to imagine themselves on Mars and bask in its glory.

His scientific images looked like tourist photos. And Peter, betting that scientists wouldn't be the only ones who wanted to look at them, made a secret handshake deal to thwart NASA protocol and post the images on the Internet as they came down from Mars. It was arguably the first ever viral marketing campaign—undoubtedly the first for space. The traffic he brought to NASA's site nearly crashed the whole Internet. The coolness factor re-awakened a waning interest in not just the Red Planet, but space exploration itself.

This is Peter's whole *raison d'etre*, as well as his gift of empathy—a rare trait among scientific minds: obsessed with discovery, but never forgetting to stop to smell the roses. Peter wants people to care about space and science, so he does everything possible to make it romantic and within arm's length. Get through that gruff exterior, and I'm just positive we'll find an old softy.

NOW PETER HAS TAKEN ON SOMETHING BIGGER. HE DIDN'T JUST BUILD the cameras for this mission, he's the captain of this whole ship and he won't take no jive from no one—except NASA. They control his $420 million budget and can cut him off at any moment, if he goes rogue. Not that I'm implying he would ever hijack a Mars lander.

Peter Smith invited me to his rocket-ship-shaped home—a design rendered when he was a swinging space bachelor—because he wanted to revive the great space narrative, begun a generation ago, but now in need of a new chapter. From our scant conversations before I arrived in Tucson, I learned he was looking for an outsider to join the mission and articulate to the world a story starring one lovable but tough-as-nails hero, Peter Smith, on one crazy, heroic, funtastic mission to explore the innards of another world.

This is our first face-to-face Mars accord. Peter wants someone on his mission that's not a brilliant scientist. *Check.* He's got enough headaches with 130 of those. He's looking for someone who might see Mars with a fresh approach and could write about it from a new perspective. *Check.* And there's one thing Peter can see clearly—I've got naïveté in spades.

Still, this whole project is a risk. Letting an outsider into Mission Control makes Peter's current P.R. chief nervous.

"You're a liability." she says. Then again, she used to work for the folks that make shoulder-fired Stinger missiles, cluster bombs, and the like. Transparency doesn't come naturally for her. I just have to gently remind her, this is the Martian arctic, not Afghanistan.

NASA WAS ONCE A BRASH ORGANIZATION, THEIR RANKS FILLED with half-crazed suicidal rocket jockeys and space cowboys. On the eve of the first moon landing, President Nixon prepared a speech to deliver if Neil Armstrong and Buzz Aldrin couldn't get back from the moon. Why? Because they weren't sure they *could* get back. NASA never apologized for taking that risk and the American public cheered them on with every small step for mankind. NASA gave us heroes fighting the good fight, expanding the possibilities of modern civilization. Somehow, since then, they devolved into a bunch of terrified bureaucrats. Can Peter Smith and his fanboy sidekick restore the glory days with a single trip to Mars? Hell, yes. If I fail, Mars and Peter could be relegated to obscurity forever and have their planetary street cred stripped away. Remember what happened to Pluto?

Peter is not going to ask NASA if it's okay for me to be here, either. That'll make life a little bit more complicated, but Mission Control is *his* building. And he decides who gets in. Besides, what's a minor security breach?

"Security breach? This is not a security breach." Peter says. "You were issued a security badge through the proper channels. We had to make sure you didn't leak stories to the press."

FOR THE LONG YEAR AFTER OUR FIRST MEETING IN TUCSON, I TRAINED. I woke up early and went to bed late. I had two full-time jobs: one paid the bills, while Mars, Peter Smith, and the Phoenix mission fueled my dreams for a better tomorrow. There was no way I would let my own ignorance keep me out of Mission Control. I did a few push-ups, but training mostly consisted of early mornings getting caught up on all things space before I headed off to my cube.

I'd give anything for a chance to spend 90 days on our red neighbor. Even though it was a gamble, Peter was never entirely sure letting an outsider in was a good idea. Or even if it was, that I was the right guy to do it. I sublet my apartment and told friends and family I was going to Mars, without a hint of irony. I even threw a bon voyage party. All my loved ones showed up to wish me luck. Then I held my breath and hoped Peter would eventually let me in.

ON THE 10TH DAY OF THE MISSION, AFTER A DRAMATIC AND SUCCESSFUL landing (Phoenix on Mars and me at the glorious Tucson Airport), the intergalactic discovery was about to begin. The engineers were satisfied that Phoenix was in good shape and the excavation could get under way. The mission held press conferences to update the few hard-core media outlets still in town. And then a ray of light shined down as Peter's assistant walked over and handed me my very own security badge. My name neatly printed in black Sharpie.

"Talk to Peter before you use this," she said.

"It's a five-day trial," Peter said. And that's where this story begins: Peter Smith offering me the chance of a lifetime, a chance that's never been offered before and one that is never likely to be offered again. I just won the space-nerd lottery and I swear I'm not going to let Peter or his team of world-class scientists and engineers down. See you in Mission Control.

CHAPTER ONE

FIRST-DAY JITTERS

SOL 11

IT'S NEAR QUITTING TIME AFTER A LONG DAY OF TAKING PHOTOS of new acquaintances—Lory and Mad Hatter—measuring atmospheric gases, and digging. The Phoenix Mars lander beeps and blips along. The sun never sets on these long Martian summer nights, but the Phoenix has strict orders to rest. The engineers want Phoenix asleep before 5:00 p.m., Mars time. Soon it will be time to put away its instruments and recharge its batteries. With the core plan nearly finished, Peter Smith and the engineers back home will be pleased.

There's just one more critical task before Phoenix can crank up its night-time heaters and initiate sleep.

"RA Acquire Sample with Rac Doc" is the instruction. This note and the corresponding lander code tells Phoenix to scrape up the first ever scoopful of Martian dirt. It's no ordinary scoop of Martian dirt. This scoop is a milestone in a long journey—one that took centuries to complete. It's the first human experiment ever done on the arctic plains of Mars. And a tiny step in the process of one day getting a man to Mars. A small camera mounted on the robot arm documents the moment for posterity.

Once this Martian dirt is safely tucked away, Phoenix will send home its daily report and then head off to bed—*to dream of finding little green men and having its day delivering a lecture to the king of Norway when it accepts its Nobel Prize on a stage in Oslo.* I know it's just a robot, but did I mention it's not coming back alive? Phoenix is a robot suicide mission.

Back on Earth, I imagine what it might be like on Mars as I swipe my security badge for the first time and walk into Mission Control.

IN CASE YOU HAVEN'T BEEN GLUED TO NASAWATCH.COM, THE PHOENIX MARS Lander is a robotic spacecraft built by NASA, the University of Arizona, the Jet Propulsion Lab (JPL), Lockheed Martin, the Canadian Space Agency, and a whole consortium of international universities and industry, all under the leadership of one intimidating scientist named Peter Smith. It carries six scientific instruments to complete its mission: find out what Mars used to be like and if anything can, or did, live there.

We don't really know all that much about what's going on down on the surface of Mars. "Is there life on Mars?" might feel like just a brilliant David Bowie lyric, but it's actually a legitimate and central question. It's worth a tiny digression to talk about the "life on Mars" issue before we get back to humanity's first interplanetary groundbreaking ceremony.

Sometimes aliens are the only things that make us care about Mars. They're the gateway drug to the hard science. There's no shame in dreaming about aliens. Even the most stone-faced scientists on the Phoenix Mars lander team imagine what might happen if they turn on the electron-scanning microscope and see tiny cells mucking about. Even better, if Phoenix found some wide-eyed E.T.s lounging on the ice, NASA would get a huge new budget, *Martian Summer* book sales would be through the roof, and Peter Smith and the Phoenix Lander would share a Nobel Prize. Win-win-win.

What's more likely is Phoenix might run into tiny bits or blobs of unrecognizables that would be hard to classify. How do you know when you have found "life" if it doesn't look like you or even anything you are even remotely familiar with that falls within your classification of "alive"? Even defining life on Earth is kind of a tricky thing. If you're too inclusive in your definition, then you end up allowing things like cars—that convert energy and move—as living. But if you're too restrictive you might designate things like mules—that convert energy and move but don't reproduce—as *not* living.

If it doesn't have eyes and use a ray-gun, and it's not a DNA or RNA or even carbon-based life form, how will you know it's alive? Finding new forms of life on another planet would send our idea of sentient supremacy into a tailspin. Since we only have one data point, life on earth, finding some strange form of life on Mars would certainly shake things up a bit. This mission does not have any sort of DNA detectors, but it's got some good tools for decoding whatever mysteries it encounters. So I'd like to quash those hopes and fears before you get too excited about this book revealing a giant Mars conspiracy of brain wave-reading Martians. But don't worry; there will be plenty of time for tinfoil hats, whether they be a fez, centurion, classic, or even a bonnet for the ladies. Read on but keep your Reynolds Wrap close at hand, because there will still be some unanswered questions at the end of the mission.

Mars was once similar to Earth. Then, a couple of billion years ago, it went from a soupy-warm planet to a cold desert. We don't know how that march toward doom happened. There are huge gaps in our knowledge: from simple things like the pH of the soil to big, earth-shatteringly revolutionary things, like is *Mars habitable*? Phoenix, the robot, is mandated to find the answers to these fundamental questions. If everything goes according to plan, we should have some answers in the next 90 days. (Or less, if you read quickly.)

Before Phoenix, remotely directed robotic spacecraft successfully reached Mars on five occasions. Viking I and II in the late 1970s, the Pathfinder in 1996, then the never-say-die rovers—still in operation—since 2003. No mission had yet ventured to the Martian arctic or brought a long shovel for digging into the surface. For a long time, planetary folk thought it was a big block of frozen carbon dioxide that rained out of the atmosphere and froze on the poles of the planet. Over time, it created huge scarps and ice sheets. Then at the University of Arizona, Bill Boynton, a smooth-headed, white-bearded scientist who races Porsches and leaves the top buttons on his collared shirts undone, discovered the north pole of Mars was loaded with hydrogen. That hydrogen was likely tangled up with oxygen in the familiar H_2O configuration (hint: it's water), suggesting that instead of tons of dry ice, there might be a giant frozen ocean below the surface of Mars. A giant ocean on Mars? That's worth looking into. Soon after, NASA selected Peter's lander project and Phoenix was born.

THERE'S A SCHEDULE POSTED FOR THE START OF SOL 11, SHIFT 1. THAT'S today: sol 11. Sols are how you count days on Mars. We'll get to why a day isn't really a sol soon but, for now, a sol is a day and a day is a sol. Today, sol 11 is not the first day of the mission (hence the reason I'm starting on sol 11 and not sol 1.) It took a week to get my security badge, so I could not get into Mission Control until sol 11. First, Peter had to decide if I was up to the task of being the official chronicler of what is essentially his entire life's work. Then they had to make my badge, and apparently that takes a bit of time as well

It's a big day for Phoenix. After ten sols of engineering check-outs, everything looks good. Phoenix is fully operational.

There won't be a ribbon cutting ceremony for the science phase of the mission (or my arrival), but hopefully there will be some ground breaking. Sadly, most of the media went home. Yesterday's televised press conference will be the last one for a while. Still, this sol is momentous.

The science phase of the mission doesn't officially start until the first experiment. Bob Bonitz from the robot arm team drops his load in Bill Boynton's TA. That's science jargon for the RA putting dirt into TEGA's thermal analyzer (TA). The TEGA oven then begins to decode Mars by sniffing out the various compounds that make up Martian dirt. In order for us to get to this one moment, a lot had to come together just so. Dr. Wernher Magnus Maximilian Freiherr von Braun had to successfully invent modern rocketry. Einstein had to discover relativity. And the parents of Peter Hollingsworth Smith had to inspire him with some fantastic tales about reaching for the stars. The rest is details.

ON MAY 25TH, 2008, TEN MONTHS AFTER LAUNCH, THE PHOENIX MARS Lander touched down on the arctic plains of the Martian north pole. The University of Arizona-Tucson is the host of this mission. It's all happening at a big warehouse facility called the Science Operation Center. The SOC is Mission Control.

The SOC is not a typical Mission Control. Since the first rocket launch back in 1950, Mission Control has been in Houston, Cape Canaveral or Pasadena—one of the NASA satellites. Luckily for Tucson (the city), Phoenix (the robot), and me (the human), this is not a typical mission. NASA tried something new. Peter Smith won a competitive bid after pitching the idea for this mission to NASA administrators. The University of Arizona, where Peter works, decided they would champion the cause and build Peter his own Mission Control once NASA gave the okay. Over the last few decades, the university developed its own mini space program. Hosting a Mars mission is a big break, a chance at real space legitimacy and the opportunity to become one of the leading space research facilities in the world.

Peter was born and raised in Tucson. His father held a position at the UA—which he took after inventing the Yellow Fever Vaccine in Brazil and saving more than 100 million lives. Try living up to *those* expectations. The university took an old Archaeology building and outfitted it with state-of-the-art communications gear and built a little Mars sound stage. The University of Arizona-Tucson wants to make its mark as world-class ultra-premium space university while their native son leads the first successful freelance Mars mission. The SOC sits on the edge of an old Tucson neighborhood among the tall saguaro cacti, adobe homes, and some impressively seedy bars.

LOOKING AT THE MOSTLY EMPTY DESKS AND GLOWING MONITORS GIVES me a chance to think about how crazy it is that I'm here. I'm about to spend my summer (almost) living on Mars. As a point of journalistic integrity, you should know that I don't understand most of what's happening around me here in Mission Control. I spent the last year learning everything possible, and that proved the bare minimum needed to not embarrass myself—at least not like the journalist who made the unfortunate mistake of asking Peter—during a press conference—what would happen to the astronauts at the end of the mission. (There are no astronauts, everyone is safe.)

TODAY, IN MISSION CONTROL WE ARE 11 SOLS INTO A 90-SOL MISSION. Prints taken from the lander's stereo camera are tacked to the walls. There are hints of possibility hidden in those images. Flashes of Mars porn titillating the minds of the bio-curious. These are the first ever images of Mars' northern plains.

They tell a story equally fascinating to the one that follows. Maybe instead of sneaking me into Mission Control, Peter Smith should have hired a famous curator to do a Metropolitan Museum of Art show about abstract Mars landscape photography. Fortunately for me, he chose a different path for his Martian story. To those unaccustomed to looking at Mars images, one thing is striking: they look just like deserts on Earth. These images are from the arctic regions of Mars and the parallels are amazing.

Peter took his team on a Mars analogue expedition to Antarctica last summer. Mars scientists are always looking for places on Earth where they can camp out and play Mars for a few weeks. The dry valleys of Antarctica are some of their favorite playgrounds. The data can help calculate how difficult it might be to acquire soil or what strange and extreme forms of life can survive in these harsh climes.

THE PHOENIX TEAM EXAMINED MARS-LIKE ENVIRONMENTS ON EARTH to improve the choices they would make during their Mars experiments—a dress rehearsal of sorts. Not perfect, but better than nothing. For instance, if you stroll through Mission Control and look closely at the images of the landscape tacked to the wall, you might notice row after row of gently rolling polygons crisscrossing through the otherwise bleak, rock-strewn vista.

These polygons are, collectively, a signal that this mission just struck gold: Martian ice. Was Bill Boynton right about a giant frozen ocean trapped just a few centimeters below the surface? If his deduction proves true, Phoenix will scrape up this million-year-old permafrost with its robot arm and start to decode the planet's history. The polygons we see on the wall seem just like the ones Peter studied on his trip to Antarctica.

"We have to be patient for results, but I just can't believe we landed on such a perfect spot," Peter says after touchdown.

The polygons on Earth form when the ice beneath them expands and contracts. It's a warping effect from dust falling into the spaces created by the retreating ice. When the ice shrinks, cracks start to form. Bits of dirt are blown into the cracks and then temperatures drop and the ice refreezes and expands. When temperatures drop and the ice expands, there's no room for the ice and warping occurs. Repeat over a couple hundred million years, and you get lovely rolling hummocks and troughs. Not as pretty as the rolling hills of Tuscany, but nicer than the pavement at the abandoned drive-in over on Roscoe Boulevard.

THE PHOENIX MARS LANDER IS NOT GOING TO WIN ANY ROBOT BEAUTY pageants. Phoenix looks like a bloated, stationary Johnny 5 with a touch of Fetal Robot Alcohol Syndrome. Her scientific guts are all exposed on a bare three-legged scaffold of a body, yet her chunky metal cylinder of a head is where most of her magic happens. But it's what's on the inside that counts. And yet, the computer inside Phoenix isn't even all that sophisticated. The smartphone in your pocket can do far more FLOPS (floating point operations per second) than the RAD6000 chip that runs Phoenix. Not that this machine isn't a wonder of modern engineering; it is. When Lockheed Martin and JPL built the original incarnation of the Phoenix body in the late 1990s, this was top-quality space hardware. It's just more difficult than you think to upgrade space-certified hardware. Space certification requires more than the stamp of a Notary Public—and it costs millions of dollars. When you add new parts to a lander there are loads of hidden costs. Since Team Peter already blew through its cost-cap by about $30 million just to make this puppy fly, there was no room for bells or whistles. Imagine Peter's embarrassment at the Explorer's club when he had to explain why there wasn't going to be an anemometer on board and the flash memory was limited to 100MB. Oh, dear. I hear they couldn't even afford touch sensors for the robot arm!?

Phoenix is a "green" lander. Sorta. Its body is recycled from the unused twin of Mars Polar Lander that crashed in 1999. That's why it's called Phoenix. It's rising from the ashes; rebirth out of the ruins (*not* because we're in Arizona). Peter liked the poetry of it all, "From ye ashes thy spacecraft shall riseth and seeketh thine Martian truth, and we shall call you Phoenix." It's what I imagine Peter said when he got the call from NASA telling him his mission had been accepted. But that might give you the wrong impression that Peter speaks like Jesus. He rarely does. His voice is more of a halting swagger with a hint of gravel and some avuncular overtones that are particularly notable when Peter explains some bit of science that's captured his—and inevitably your—imagination.

The imagineering done on Phoenix comes courtesy of the science payload—the scientific machinery it carries. This payload consists of six instruments designed to characterize the properties and makeup of the Martian environment. The Phoenix will use its robotic arm to dig up soil samples and run experiments to determine what's in the ground and how it got there. The instruments range in complexity from a simple wind-measuring telltale to an extremely sensitive atomic force microscope. Some of the friendly scientists you're about to meet described their instruments to help you better understand the little friend you will follow throughout these pages. The glazed-over mind-wandering feeling you get from reading about these instruments is just your brain making the leap to hyper speed. Don't be alarmed.

Robotic Arm (RA)

The Phoenix Mars Lander is a *dig and eat* mission. In order to be a success, the lander has to break ground and scoop up some dirt. So they need a long digging arm for the task. The robot arm (RA) is just under eight feet long with an elbow joint in the middle and a bucket scoop on the end. Sticking out from the backside of the scoop is a circular rasp. Its job will be to acquire icy-soil samples if the team is lucky enough to find them.

The arm reaches out over the lander, scoops up Mars dirt and then dumps it into the other science instruments. Needless to say, the RA engineers are all really good at that game where you try to pick up the stuffed animal with the claw. I suspect this talent was a critical factor in how they got their jobs in the first place.

If you want to dig into the permafrost on Mars, why not bring an ice-coring machine instead of a digging arm? Funny I should ask. It is a good idea, a planetary scientist's dream to be exact. The problem with bringing something like a huge coring device is, simply, weight. It's simply difficult and expensive to get heavy things to Mars. We can estimate about five figures per pound of weight we bring along. You think the airlines baggage fees are excessive until you pay NASA an extra couple hundred grand for your carryon. Shovels and drills are far too heavy to bring on a budget-conscious mission like Phoenix, and so a delicate robot arm must be precisely engineered to perform a wide array of tasks with its little (and light) claw.

Robotic Arm Camera (RAC)

The RAC is attached to the RA just above the scoop. The instrument provides close-up, full-frontal color images of the Martian surface close to the ground, under the lander, or anywhere the RA can go. Its got all kinds of filters and scientific attachments to capture and make sense of extreme close-ups of dirt or whatever else Phoenix can dig up. I for one am hopeful for a secret decoder ring.

Surface Stereo Imager (SSI)

The SSI functions as the eyes of Phoenix. It takes the pretty postcard pictures you might see on the Internet. The design is based on Peter's famous stereo-imager built for the Pathfinder mission. That was the first Mars camera to use a Charge Coupled Device (CCD) like you'd find on your digital camera at home. Since then, it's had a few upgrades but it's still your classic Mars imager. For eight to ten million space bucks, Peter will build you one too. The SSI has precisely manufactured glass lenses and flawless resolution. Situated atop a large mast, SSI will provide images at a height of up to two meters above the ground, roughly the height of a "tall" person. The two lenses on SSI simulate the human eye, creating three-dimensional stereo vision. It's loaded with all kinds of filters to create images in various regions of the light spectrum. These filters will help the team figure out what they're looking at, whether the reflective object they see might be ice or just some shiny bits of rock.

Microscopy, Electrochemistry, and Conductivity Analyzer (MECA)

MECA is made up of four instruments: a wet chemistry lab, two micro-scopes, and a conductivity probe. The first real chemistry work done on Mars involves dissolving small amounts of Mars dirt in water—brought from Earth—with the unironically-named wet chemistry lab (WCL) to determine the pH, what types of minerals are present, and their conductivity. MECA contains four single-use WCL beakers, each of which accepts one sample of Martian mud. Phoenix's RA will deliver a small sample to a beaker, then a pre-warmed and calibrated soaking solution is added. The optical and atomic-force microscopes complement MECA's wet chemistry experiments. With images from these microscopes, scientists will examine the fine detail structure of soil and water ice samples. Who *knows* what else they might see? MECA's thermal and electrical conductivity probe is attached at the scoop joint, where the scoop meets the arm of the RA. The probe has four small spikes that can be pressed directly into the ground. And this probe can read temperature and humidity of the air and measure the temperature and conductivity of the soil.

Thermal Evolved Gas Analyzer (TEGA)

TEGA is a combination high-temperature furnace and mass spectrom-eter instrument used to analyze Martian ice and soil samples. Basically, it's a robotic nose. The instrument drives off gas that the sensors inside can "sniff" at various temperatures. Magic happens. It works like this: the robotic arm delivers samples to a hopper designed to feed a small amount of dirt and ice into eight tiny ovens, each one intended for a single use. Once received and sealed in an oven, the sample cooking begins. The engineers carefully increase the temperature at a constant rate, and closely monitor the power needed to heat the sample. This pro-cess, called "scanning calorimetry," shows the transition of the sample as it decomposes into its gassy components. The gas that's released is passed on to the mass spectrometer for analysis. This information is vital to understanding the chemical makeup of the soil and ice.

Meteorological Station (MET)

MET will record the daily weather of the Martian northern plains. Using temperature and pressure sensors and a crazy first-time-on-Mars laser beam light detection and ranging (LIDAR) instrument, MET watches the weather. Every lander worth its salt should have a laser beam. The MET's LIDAR works sort of like RADAR, using powerful laser light pulses rather than radio waves. The LIDAR transmits light vertically into the atmosphere. This light is then reflected off dust and ice particles. The instrument collects and analyzes the light to reveal information about the size of atmospheric particles and their location. MET provides information on the current state of the polar atmosphere as well as how water is cycled between the solid and gas phases in the Martian arctic.

These aren't the most sophisticated instruments technology has to offer. They are simply some of the most sophisticated instruments you can safely get to Mars for under a billion dollars. They were carefully designed and tested to squeeze stellar results from a meager space budget and countless restrictions. Constructing instruments for Mars presents all kinds of challenges that terrestrial technology simply doesn't have to deal with. It's a challenge that's hard to appreciate until highly calibrated sensors give odd readings or valves freeze open 200 million miles from home. Sensitive lab gear is never meant to be strapped on the cone of a missile, irradiated for months on end, and then slammed onto the surface of a dusty cold place. It's meant for a sterile, temperature-controlled lab and gentlemen who wear white coats.

THE CLOCK ON THE WALL IN THE SOC READS 13:56 LOCAL SOLAR TIME, Mars. It's 22:30 Local Tucson Time, Earth. This is it! My first day on Mars. I hope the kids are nice and no one steals my milk money. Each sol starts with a kickoff meeting, a short assessment of what's happening with Phoenix that given day. There are quick updates from key team members on the lander's status. The team reviews the plan the Phoenix just finished on Mars and looks for issues with the instruments, data transfers, the weather, and any general health or safety concerns. Today, it's time to talk about our very first experiment. The scoop.

Scientists file in, wish each other "Good morning," "Good afternoon," or "Good evening." Basically no one has any clue what the proper office banter should be. All bets are off at the water cooler, too. It's innocuous social norm chaos. Even though it's night on Earth, it's early afternoon on Mars. And for those of us still getting used to Mars time, we're just starting our days. It's confusing and disorienting. Still it's charming to see people eat ice cream—as many are today—and offer perfunctory greetings. (Free ice cream—and not the dry chalky astronaut variety kiddies whine for at space museums and then kick to the curb when they realize that freeze drying actually removes all the great things about ice cream, namely ice and cream—is one of the perks of life in Mission Control. Truth be told, ice cream practically fuels the mission. And has done so on past Mars explorations. Why? The tradition remains one of the true Mars mysteries.) Chit-chat turns quickly to "the scoop." *Not* the ice cream variety. It's time to get on my eavesdropping ears.

Standing at the front of the room for today's sol kickoff is Joel Krajewski, the Surface Phase leader for the mission. Peter ambles into the room. He looks like he just woke up.

"Good morning," Joel says to Peter and the assembled team. Joel is in charge of systems engineering for the surface phase of the Mars mission. Let me tell you, he's not your average Joel. He has a very engaging demeanor. When you're one-on-one, it's like you're talking to the engineer version of a Hollywood producer. He's got the salt and pepper hair, closely cropped beard, and height to match L.A.'s slickest. Just replace the Prada loafers with socks and sandals. He might have been awake for 30 hours and currently be thinking about how to save Phoenix from impending doom, but he makes you feel like what you're saying is really important. It's a good quality to have when you have to herd sleep-deprived scientists all day. He is part of the Jet Propulsion Lab's senior mission management team. He developed the process that brought together and trained this consortium of brilliant scientific minds.

Krajewski's team must inspire the engineers on Phoenix to execute the science vision without destroying the lander. He's the underwriter for scientific harmony. Getting team members from government, industry, and academia to play nice is crucial to the success of the mission. There is one overriding and constant source of tension: engineers who want to protect the spacecraft pitted against scientists who want to push the limits and see what the lander can do. This is a kamikaze robot mission. In training for the mission, Joel and his team had to figure out how they could get the team to eschew conflict and harness this creative tension to get the work done.

Putting everyone under near-impossible conditions of stress so they bond over the common hardship is a staple Jet Propulsion Lab management technique. Why not try it at your next shareholder meeting? Krajewski's team of systems engineers has spent years thinking about just these sorts of Martian planning problems. I won't pop-psychologize at this point, but rest assured I'm making mental notes about the effectiveness of this exercise in promoting team chemistry (no pun intended).

PLANNING COMPLICATED TASKS IS ALWAYS HARD. MARS PLANNING IS particularly complicated because Phoenix does not operate in real time.

"The long delay complicates things," Joel says. A sol-long feedback loop turns the simple act of digging some dirt into a daily drama. Scientists and engineers act out this epic in several acts over the course of 15-18 hour sols. It's the weirdest reality series you will never see on TV—a bizarre mash-up of *NOVA* and *The Office* produced by an army of scientists and engineers.

Mars is so far away that direct control of the lander is, for all practical purposes, impossible. Here's why:

It takes about 15 minutes for a message to reach Mars from Earth. Traveling at the speed of light (299,792,458 m/s), it goes to Mars, which, depending on where it is in orbit, is about 75,000,000 to 375,000,000 kilometers away. Now do some math: *Divide, carry the one, convert to minutes, and voila! 15 minutes.*

"You can think about it taking eight light minutes for a message to travel 150,000,000 kilometers—that's about the distance to the sun. Mars is about half as far again. Well, a little less. So that's 15 minutes," Peter says. The signal has to come back too; so multiply by two and you get 30 minutes round trip. That's a long time to wait when you need to talk with your lander *right now*. Try playing *Dig Dug* on a thirty-minute delay and you'll realize that it is impossible. Instead of trying this extreme Atari approach, Mission Control engineers rely on uploading an entire day's sequence and then waiting for the lander to give it a go and report back. The lander sends reports and incremental status updates through a satellite relay.

"Don't forget that the satellite relay adds another one to two hours delay; that really kills us," Peter says while he draws me a picture of Earth's orbit and a message-relay diagram. "Phoenix can't message the Earth directly."

Planning the lander's whole day in advance is a tough pill to swallow for even the most mission-hardened engineers. Unending planning debates are often at the heart of mission dyspepsia. Since almost everything on Mars falls under the category of "First Time Ever," you enter a lot of situations blind and grasp for something familiar to help you analyze the situation and create a plan for your waiting lander. The lander needs a steady diet of plans, one a day, day after day. Often there's just not enough time to really understand the situation before moving on to the plan for the next day. That explains the current structure of planning and hoping. You plan for a sol. Wait. Hope you made good choices. Get results. Re-plan. Re-hope. Eventually you start to understand.

The zig-zagging uneven nature of planning can make it hard to stay connected to what day it is when it takes so long to get results and then adjust your course. I would liken it to the dissociative effects of taking too much cough syrup. It seems like the right idea but it's really disorienting (and not that much fun).

You don't just show up at a Mars mission and go to work. Organizing the personnel and procedures for conducting planetary expeditions requires years of training. There should probably be an entire Columbia Business School curriculum on Mars management. (And Joel K. should run it.) Joel and his team wrote the "How to Operate your New Mars Lander" Manual and then trained the scientists how best to use their new toy. What should be a two-year course was abbreviated for busy scientists. Still, the training missions they *did* provide in the year before launch were just enough to make it work. These training/test sessions were required by NASA to show that it was possible to operate Phoenix effectively.

"When we first built the planning system, it took us about two months to execute a single sol's worth of work," Joel says. "By landing day, they shaved it down from about 200:1 to 1:1—well, on most days."

Getting brilliant and detail-oriented planetary scientists to work together and quickly is a constant challenge. Even with training sessions, the daily schedule requires an impossible amount of work. Joel resorted to tricks in order to keep pushing forward in spite of uncertainty and imperfect information. There's no *Kobayashi Maru*, as fans of Captain James T. Kirk might imagine. Still there is some pretty brilliant stuff, like the 80% rule. This clever ruse offers the detail-oriented engineer a way to bypass his or her own perfection bias. If an engineer has a complicated task—digging to a precise depth or taking a difficult series of photos—and completing the task seems probable but not necessarily certain, they're supposed to invoke the 80% rule. The rule lets engineers say they are willing to proceed with 80% certainty. They don't have to put their credibility on the line, just be reasonably sure they can complete a task. Why would you need something like that? Because all these hyper-smart people tend to over-scrutinize. Achieving perfection in the face of confusing or conflicting results can take too long. When you don't have a lot of time, but a lot of difficult operations, this is a problem. It can cripple your mission.

"Whenever they ran into some uncertainty, they'd take too long before coming up with a new plan," Joel says. With the 80% rule engineers give themselves permission to be wrong—even though they're mostly right. "It was key in speeding up a timeline that got bogged down in the desire to do things perfectly."

ONCE HE WELCOMES EVERYONE BACK TO WORK ON MARS, JOEL HANDS things over to the engineer seated to his right. He's tall and unshaven. He looks a little rumpled sitting in the too-small plastic folding chair. The chair is labeled "TDL, Tactical Downlink Lead." There are 18 such folding chairs, all with labels like TDL, or MET, or CSTG. Acronyms are a Mission Control dialect. First, you despise them for their obfuscation, then they're cute and funny. A tactical downlink lead (TDL) tracks the millions of data packets Phoenix sends to Earth each day. If something goes wrong, or data packets that carry Phoenix's instructions are missing, it's the TDL's job to track them down.

Today's TDL, the tall guy, squished into his folding chair, is an engineer named Jim Chase.

"All data was returned, although some non-critical data will need to be re-transmitted," Chase says as part of his comprehensive and highly technical data report. He fumbles with his papers as he talks about what data goes where.

Seated near Peter at the opposite end of the conference table in the downlink room of Mission Control is the representative from the spacecraft team. He gives his morning report. It's a kind of nonsensical list of acronyms that I hurriedly look up in my acronym dictionary. (Yes, this exists, and I love it.) This is genuine space talk. No flim-flam. It's the talk we came for.

"Blah blah subsystems . . . channelized telemetry . . . something . . . block validation . . . ISAs . . . ISEs," the spacecraft man says. Even if we can't understand a thing, it's perfect. At the end of the engineer's report there's one statement in plain English.

"The spacecraft is healthy," he says. Phew.

"Is there a weather report?" Joel asks. Palle Gunnlaugsson, an Icelander on the atmospheric team, is today's Mars weatherman. He hasn't got any smiling suns or high and low pressure systems moving in from the north. That's not really Palle's style. He's a young, corpulent Viking who hangs out in dive bars and chain-smokes. He also hand-wove the Kevlar strands on the Phoenix telltale, similar to the fluttering strings you might find on a sailboat, and used on Mars as a clever wind-measuring device. It's comforting to know we've got Viking tradition with us on our trek. Palle reports that everything is the "same as yesterday."

"How was that?" asks Krajewski.

"Same as the day before," Palle says with a faint smile. Icelandic humor, like the Mars climate, is very dry.

"Okay, let's break. Data will flow shortly. I'll see you all back at midpoint," Joel says. Once dismissed from kickoff, it's time to review tomorrow's activities. There's a little guesswork involved. The team works around the assumption that yestersol—that's not a joke, that's really how they say yesterday on Mars—Phoenix dug its first little trench and scooped up a sample. For all our scoop and sample talk, there is no hard evidence that "RA Acquire Sample with RAC Doc" actually worked. We must wait until the data downlink; this is when the process where we download data from Mars (and conveniently, the name of the room we stand in) begins. Any moment now the lander will call home and send its bits through the DSN (Deep Space Network) and into the SOC. Then images from the SSI and RAC will tell us if our dirt is safely tucked away in the back of the RA scoop.

"Please look out for scoop images," Joel says as the teams head off to their offices and desks. As if he has to remind them.

Break time. Now I'm supposed to put on my fedora with a badge that says "cub reporter" and unlock the secrets of the mission.

Unfortunately, everyone seems to disappear the minute the meeting ends. Or put their noses down at their desks, making it clear this isn't question time. There doesn't seem to be a hot scoop to be found anywhere, so I get some coffee and read some Martian literature in the kitchen.

LIVING IN MISSION CONTROL IS *EVEN MORE* BIZARRE THAN YOU MIGHT think. Both the mechanics of Mars and the extreme delay communicating with the lander contribute to a wacky sol schedule. One of the quirks of the Mars planning schedule is that it means working on Mars time. Mars does not have an ordinary 24-hour day. It has an extraordinary 24 hour, 40 minute day. We work the Mars night shift coming into Mission Control just as the day ends on Mars. The scientists build a plan while the lander "sleeps" and then upload it before the new sol.

What time does lunch start if you begin work at midnight? Today, I brought a cheese-and-mustard sandwich for lunch. But if I eat my sandwich now, it could be tough to find food at 3:00 a.m. if I have to—gasp—step out of Mission Control. The whole idea of living on Mars time is ludicrous. Working on Mars time isn't like being in a different time zone. No, it's much worse.

First, a little background. A "day" is how long a planet takes to complete a rotation on its own axis. For the Earth, that's 24 hours (by definition). Mars, however, spins a little slower, which means it takes slightly longer to complete the rotation. Consequently, Mars does not have "days." You already knew that, though. Here we have our "sols," short for solar day. The solar day is 24 hours, 39 minutes, and

35 seconds—about 39.5 minutes longer than an Earth day. This isn't a concern for most Earthlings, unless you're part of the .0000022% of the population who want to work nights on Mars.

If you're on Earth and you change time zones, it upsets your circadian rhythms and you feel miserable. Jet lag isn't just an annoyance from sitting on a plane. There are physiological effects from monkeying with your circadian rhythms. These rhythms regulate hunger, sleepiness, hormone release, and body temperature. When out of synch with each other and the sun, you feel jet lag. When you fly across the Atlantic, your sleep is disrupted and you feel a little groggy while your body realigns itself. What if your body can't readjust? Working on Mars time is essentially like changing time zones every day. You can never adjust; and you never get used to it. Instead you fall farther and farther into a sleep deficit. Your changing sleep cycle results in physiological changes; the hunger and hormone cycles go cockeyed and madness ensues. In the short term, you get cranky; after a month or so, your endocrine system starts to misfire, and, if you keep pressing past these stages, you go insane. It's bad for you to live on Mars time. Thankfully, we'll only subject ourselves to it for three months. Any longer and *Martian Summer* would end like *The Shining*.

In order to keep track of the Mars day, you have to map a 24-hour, 39-minute clock to a 24-hour Earth clock. Feel free to use a calculator for this section. You need to shave 40 minutes off each day to keep in synch. Practically speaking, if today you're operating on Mars time and go to work at 10:00 a.m. Pacific Standard Time (PST), tomorrow you'll go at 10:39:30 a.m. PST. Every 36 days, you're back at 10:00 a.m. PST—but you've lost an entire day in the process.

It's like the old adage "Fall back, *Mars* ahead . . . 39.5 minutes." You're not familiar? To keep you on track, there are several clocks in the SOC that denote Mars time (Local Solar), Local time (Tucson Standard), and UTC (Coordinated Universal Time, like Greenwich Mean Time for Intergalactic Scientists).

While you gain forty minutes to sleep in every day, you get to spend them feeling paradoxically sleep-deprived and alternately apathetic and/or anxious. One fun outcome occurs when your endocrine system starts to sputter from perpetual jet lag. The flow of adrenaline and other hormones is stunted and the world around you becomes gray and monotone. You care less and react slower. It's a condition described to me by one of the psychologists from the counter-fatigue group as "flat affect."

"Oh . . . looks like my house is on fire," Edna Fiedler drones in a monotone to show me what's to come. There's a whole group here studying us guinea pigs for these effects. They'll monitor our descent into madness with the hope they can prevent future Mars scientists from succumbing to the same fate. For research purposes, many of the scientists opted to pee in a cup every day. So if you see lots of folks carrying their own urine around, don't be alarmed. That's not part of the madness, it's just science.

One of the NASA shrinks looking after us is Walter Sipes. He is responsible for the mental health of NASA astronauts on the International Space Station. Sipes and his wife, Edna Fiedler, a Harvard psychologist, help those working on Mars time not go nuts and destroy the spacecraft. I'm curious if this will cause any of us to drive cross-country in diaper-wearing revenge-fueled rage. (If you're not familiar with that astronaut adventure, you might want to Google it.) In an effort to appear civil and professional, I don't ask Walter or Edna about that particular incident, but I regret it.

Apart from feeling crappy, it's really annoying to keep track of time in Mission Control. If you want to perform anywhere close to normal, you have to delay the effects of Mars time by adhering to the Mars clock as best you can. Switching from Earth to Mars time will hasten your lunacy and is not recommended. To keep track of the Mars day requires a Mars clock. Your Mars clock will map Mars time from a 24 hour, 40 minute day to the standard 24-hour day. A Mars clock needs to lose 40 minutes every 24 hours. Unfortunately, these watches don't come in your official NASA welcome basket. Sadly, there isn't even an official NASA welcome basket.

I was hoping Rolex would sponsor the mission and hand out Mars watches to the junior team, but no such luck. But there's a master watchmaker in Montrose, California, who will make and stamp a genuine Mars watch just for you for $250 plus tax.

"I only made 1,000 so it would be a collector's item, but I wanted it to be reasonable," he says. Could be worse. A few engineers at JPL asked the master, Garu Anselarian, a Lebanese man with a sweet round face, if he would make a watch that could keep track of a 24 hour, 40 minute day. They told him it would make their lives easier when they worked on a mission and that he'd also be able to sell them as souvenirs.

"I ruined a lot of watches looking for a solution," Garu says. "I tried adding gears, lengthening the hair string," but it didn't work. After a few months of tinkering he came up with a solution, a flash of genius.

"I realized I could add weight to the balance wheel," he says. He put a precisely measured dab of silver solder on the second hand spring to slow down the timepiece; a masterstroke in Mars time conversion. He built the originals from the innards of a regular old Citizen quartz watch and put a little picture of Mars on the face. (Unfortunately, sales of his masterpieces were, just like the Mars watches, a little slow.)

"For me it was a success, not much money in the Mars watch business, but I'm happy," Garu says.

Before I knew what it meant to go to Mars, I pictured myself sleeping in a little tent next to the engineering model of the Phoenix Lander. Each morning I'd wake with the one of the engineers sounding reveille and set my NASA-issued Mars watch for a brand new sol. No one plays reveille in Mission Control, and I had to make a pilgrimage to Montrose, California, to buy my own damn Mars watch.

WHEN I COME BACK, THERE'S A CROWD GATHERED IN THE SSI OFFICE. Pat Woida, an SSI engineer, is wearing his Mission Control uniform: a Hawaiian shirt, sweat shorts, and Birkenstocks. He's more excited than usual. "We got it! We got a sample!" Everyone goes nuts. They clap and cheer. Elbows push me aside to get the best scientific angle.

Pat Woida is larger-than-life—both in character *and* appearance. He is the red-haired Santa Claus of space—always jovial, spreading Martian cheer wherever he goes, and bringing gifts of freshly printed Mars images. Pat's in charge of the large-format printer in the SOC. Mark Lemmon, the co-investigator for SSI, Woida's boss, looks over his shoulder. Lemmon—who wears khakis and polos and always has his dark hair neatly parted—couldn't offer more of a contrast when standing next to Woida. Yet both men can barely contain their mutually shared joy.

"Wow. Look at that," Pat says and points to the images on his machine. They look at a close-up image of clumpy Martian dirt that fills the scoop. The gathering crowd spills out of the SSI office. Those who can't fit inside the office crowd around a large flat-screen display labeled "Jedi."

"I remember when the first images of Viking I came back from Mars," Pat says. He was just a pimple-faced teen when we first landed on the Red Planet, but it changed his life forever. Working on Mars is a childhood dream-come-true for Pat and almost everyone else crowded into the small office.

"The first images ever taken on Mars were of the Viking lander's footpad. It was really Mars! And then I thought one day, I'm going to be that guy," Pat says.

Cut to the present. Pat took the images of Mars we see in front of us. Dream fulfilled. Last August, he left launch to deliver the keynote address at the Star Trek convention in Las Vegas.

"That was definitely a career highlight," Pat says. "The fans were awesome. I really think people will want to know about this mission."

Outside the SSI office, Matt Robinson, the young RA engineer who wrote the code for the "RA Acquire Sample" activity, celebrates.

"We really got it," he says with a big fist pump. He started working on Phoenix fresh out of grad school. Since then, every single day for the last five years, he practiced for this moment. That's a lot of practice scoops. Now, he's grinning ear-to-ear. You just want to hug him. Happiness is infectious. He puts his hands on his head and does a kind of post-Superbowl interview.

"Five years I've been scraping up concrete. And now we're on Mars," Matt says and punctuates his cheer with another fist pump.

"Good work, Matt," Ray Arvidson, the understated co-investigator for the Robot Arm—code-named "Dig Czar"—says to congratulate the young engineer. He gives him a firm handshake and a close-lipped smile and nod.

"We haven't looked through the data yet, but everything appears to be just fine," Matt replies. He's more serious and composes himself when addressing Ray.

Peter sees the commotion from his office and joins the celebration.

"Coming through," Pat Woida says as he grabs an image of the scoop—the Phoenix Lander's first official soil sample—off the printer.

"Sign here," Pat says as he hands the image and a Sharpie to one of the RA engineers. When the RA team finishes signing, Pat proudly tacks the image to the wall.

"I wonder if there are ice crystals in there," a scientist says to a colleague as they walk through downlink. There's wide-eyed speculation and commentary coming from everywhere.

"When I look at that image I think about the Sistine Chapel," Joel says. "You see a beautiful work of art. But I can also smell the paint and hear them putting up the scaffold." You can't see the image in isolation when you spend five years building the scaffold, or a robot arm.

"These are historic images," Peter says to me in his least intimidating voice. "We got the first sample. . . . And we're all waiting to see what it means. But there's a sense of unreality. It's a bizarre feeling because you only get to see it through pictures. It's happening 170 million miles away. You're removed from the visceral direct experience," Peter tells me.

You watch all kinds of movies or TV where people beam off to other planets and the mystery is wrapped up in an hour and it's easy to accept. There's a suspension of disbelief. Oddly enough, this is the real thing, and it's unfathomable that we are controlling an 800-pound lander so far away. And not only are we controlling it, it just sent pictures back through interplanetary broadband! That hurts my brain. And my brain rejects it as false. Mars presents the first time I ever had to suspend disbelief for something that's real. Peter probably knows this feeling better than anyone. When he tells me about his disbelief I even think I detect a hint of sadness in his voice. (Martian scientists also cry.)

"Now, the question is what's written in that dirt," Peter says, examining the new images with a few of his colleagues. I try to nudge my way in and look thoughtful.

CHAPTER TWO

THE CLOD

SOL 12

BEEP. BEEP. THAT'S THE SOUND MY SECURITY BADGE MAKES WHEN I let myself in the front door of Mission Control. Thankfully, it works. I swipe it one more time just for fun.

Excited to start the sol, I show up two hours early. It's just passed 11:00 p.m. local Tucson time, and kickoff doesn't start until 1:15 a.m. local Tucson time. The big conference room on the other side of the security door is empty. It's the calm before the swirling activity of scientists meeting at the big conference table and the hotbed of Martian discovery.

The Mars schedule gets updated each day in an email called "phx_surf_ops." I'm not on that list.

"It wouldn't be appropriate," Sara Hammond tells me. So I'm not allowed on the "phx_surf_ops" mailing list. Even though I've got a security badge, "phx_surf_ops" is an official JPL email list. They're particular about who gets their email. Since I don't exist in their system, that would be impossible. People would ask questions. And then I'd have to try to explain all over again why I'm here.

"It just can't happen," she says, driving home the point. I'll live with it. And my early arrival turns into a happy accident.

"I am Nilton Renno, from the University of Michigan!" says a bright voice from a shiny dome. He says "Mi-chi-gan" with a distinguished foreign accent that I can't put my finger on.

"I'm from Brazil," he says. I think I'm sitting at his desk. He doesn't seem too fussed, though. "Are you working with the Canadians?" Nilton Renno puts out his hand and we shake. He's smartly dressed with pressed khaki pants and a fashionable collared shirt—more of a Silicon Valley executive look than Mission Control.

"Actually, I'm helping Peter with a special project, trying to bring the mission to a larger audience," I tell Nilton, over-sharing just a little because of this dogged space inferiority complex and my desperate need for credibility. Luckily, Nilton knows no over-sharing.

"I will show you something," he says, taking a seat at his desk. Nilton calls up an image from the Phoenix planning software. It's an image called Snow Queen.

"This is my favorite image," he says. I like the idea that each scientist has his or her favorite image, some piece of Mars that he or she especially loves. Snow Queen is a big shiny spot under the lander. This photo captures not only the shiny spot but also exposes the nether regions of the lander, the struts and thrusters. If we were to get all arty-farty, we might say it tells the story of a Martian vastness interrupted by a highly engineered lander intruding on the landscape. While the shadow of the lander obscures parts of the scene, the sunlight peeks in to expose just enough of the smooth white feature and thruster cone for us to begin our analysis. It's both a metaphor for our presence and hard scientific evidence. Or something like that.

"This says ice to me." Nilton says, focusing my attention under the lander and away from overreaching artistic pronouncements. "It's a hard, shiny material. That's something we predicted in the lab." Nilton is speculating at the moment, but his speculations come with years of experience and intuition. He wants to take more images and he thinks it represents everything we're here for. "Can you see these little chunks stuck to the leg of the lander?" Nilton asks.

I nod affirmatively. "The thruster plume excavated down to the ice. This is the debris that splashed up on the lander. When the lander touched down, there was a lot of force applied to the surface and that force kicked up a lot debris. I think some of it froze to the cool spots on the lander legs. That's this part here," he says, pointing to the legs of the lander. There are lots of secrets hiding in these images. Today we're hopeful that the first mission experiment, putting a soil sample in TEGA, will offer clues to understanding the shiny patch and little nodules in the Snow Queen image. Nilton tells me it's going to be a slow reveal and to be patient. I'm glad I got here early.

MISSION CONTROL STARTS SHOWING SIGNS OF LIFE. SCIENTISTS AND engineers exchange chipper "good mornings" and the overhead lights flicker on. In spite of the bags under their eyes, everyone seems excited to be coming to work.

"How's the LIDAR working, Jim?"

"Good! And your AFM scans, Tom?"

Apart from no one knowing exactly what time of day to reference, it's like any other place of work, only the bullshitting around the SOC has a fun Martian twist. That and our day starts at 1:00 a.m. local Tucson time, then some other time tomorrow and another time after that.

The engineers set down their satchels in the rows of offices that surround the central meeting area, downlink. "Downlink" is a popular word around here because it's the room where we spend most of our days planning Phoenix's adventures; downlink is also the most exciting part of the day, when we get data from Mars. The downlink process happens in the downlink room. Along the back wall of the room each instrument team has an office, MECA, MET, TEGA, SSI, RA, and RAC. Then there are offices for data managers, mission managers, and the image processing. Opposite these offices, Peter Smith has a large office, and he's flanked by the mission's accountant and the SOC general manager, Chris Shinohara.

After collecting their thoughts and completing their morning reports, the team gathers around the big decision-making conference table. Peter pokes his head out of his office and makes his way to his seat at the head of the conference table.

"I have to balance Mars time, NASA time, and the press," Peter says with a yawn. He's getting less sleep than most. With another 80+ sols to go, he already looks like a bear fighting off hibernation. I hope he makes it. His assistant brings him a large cup of coffee. He takes a big gulp and stretches.

"We're working on a big four-sol plan," Ray Arvidson says. This includes the preparation and execution of the first TEGA experiment.

"First step is to get the dirt into the oven. Then confirm the delivery and start the bake," Ray explains. The TEGA bake will take a few days. It comes in three stages, a low, medium, and, finally, high temperature experiment that will reach 1000 degrees Celsius.

"There's certainly some white material in Dodo," a scientist says, referring to some whitish chunk of what might be ice or salts at the bottom of our new trench, called Dodo. Why is it called Dodo?

"It's from *Alice's Adventure in Wonderland*," Peter says. He wanted the names to appeal to kids, so they named them after fairytales. "On Pathfinder we named them after cartoon characters, but NASA worried we would get sued for copyright infringement. I don't think we ever got sued. Although, I think the Cartoon Network sent us some T-shirts. This time we chose characters in the public domain."

"It could be salt crystals," a colleague says as they fantasize about what the next few sols might bring.

Today is sol 12 and we're working on the plan for sol 13. Before the sol 13 comes together we need a progress report from Phoenix. We take our seats so that the first meeting of the sol kickoff can start. From my squatted desk, next to Nilton's, things seem pretty positive. Some of the TEGA engineers are already shaking hands in celebrations.

"We finally got our sample," one of the TEGA engineers says coolly.

"You don't know that yet," Peter says. He shakes his head. Peter wants some hard evidence from Mars before he draws any conclusion. The TEGA engineers are willing to hope a little. They came all this way, scooped up the perfect sample and couldn't get it to slide off a shovel and down TEGA's gullet.

Up until this point, the engineers conducted various checks for the "health and safety" phase of the mission. These were the first 11 sols. The science team had to wait until engineers made sure Phoenix was upright and could safely deploy its solar panels, digging arm, and camera. Then check out all the instruments for signs of any problems. Phoenix did just fall out of the sky after a ten-month catapult through deep space. It makes sense to do a comprehensive exam before getting started. During this first week, the SSI camera made a 3D map of the space around the lander. The instruments were retested and calibrated and the RA team practiced digging. Now that all the engineers are happy and Phoenix appears mostly fit. At least it's in good enough shape to try our first experiment. All that's left is the first scoopful of dirt to fall into TEGA and discovery begins.

Ray Arvidson, the Dig Czar, calls the science team to order and begins the sol kickoff meeting.

"We started the science phase of the mission!" Ray is pleased to tell the team. *He* doesn't think it's premature to celebrate either. There is applause. Peter frowns.

Perhaps yielding to Peter's prudence, Ray retreats from his proclamation. "Today's work will be to verify that we actually started the science phase of the mission. We won't turn on TEGA unless we confirm there's something in the oven." Ray then asks the TEGA engineer—speaking on behalf of the instrument team—what signals they'll be looking for to indicate that the oven is full.

"What do we need for 'go'?" he asks.

"We'll be looking for 'oven full' or 'oven not full' from the instrument software," he says. "Then we'll be looking for the oven closure. We'll look at the current drawn by the actuator." That answer satisfies Ray. (Although I suspect he already knows these answers and is just making sure everyone is clear.)

Ray dismisses the team. In 15 minutes, the satellite relay orbiting Mars will come into view of Phoenix's helix antenna. The data transmission begins and the "oven full" or "not full" signal will flow through the deep space network and back to Earth. It's important to remember that all this digging and dumping has already—or *should have* already—happened. Phoenix is on its way to bed and we're waiting for a progress report. If everything went as planned, the scoop of dirt is now resting in TEGA's TA-4 oven, waiting anxiously for the engineers on Earth to upload a plan that tells it to start cooking.

Now we wait. Smoke if you got 'em. I'm going for some Tang.

WHEN I GET BACK TO THE BIG CONFERENCE TABLE IN THE DOWNLINK room, everyone is standing around with worried, crinkly foreheads and furrowed brows. The data is slowly streaming in from the deep space network.

"No dirt," I overhear an engineer say. The science team is migrating over to the RA office. Matt Robinson—one of the RA engineers who stayed up all night making sure that nothing went wrong with the sample delivery—is sitting at his desk. He looks up at his monitor with a squinting, contorted face.

"There's nothing in there," he says. The scoop is empty. The robot arm reports from Mars that it successfully executed all its commands. But the TEGA reports from Mars indicate something different: its oven is empty. Someone must be lying. The scoop made the drop but the package never arrived. What happened? Did they miss the delivery? Did the dirt blow away?

"We don't know yet," is the answer from Matt. He explains that they took a series of images to show the lander deck in case something like this happened. Only the images haven't arrived yet. We need more images to complete the story. Currently, they're held up due to network traffic—somewhere between Arizona and Mars. We stand by.

Matt Robinson scrolls through an inconceivably large table of figures.

"There's no way we missed that," he says to no one in particular. One of the engineers asks if it's possible they stopped looking before the sample was delivered. He means that they were measuring light from the LED and photocell at the wrong time. If the timing was off, they might have turned the "oven full" detector on too soon. This would mean that there could be dirt in there. But it would still read "oven empty" in their data set.

Good theory.

"I don't think so," Bill says, shaking his head.

Bad theory.

They look at the LED readings again. If there was dirt in there, the light sensor would be at least partially blocked and have a lower light reading than normal. Then they might guess the oven was in fact full.

Peter and a few senior engineers from JPL grab some folding chairs. They make camp in front of the big flat screen labeled "Jedi." This area in front of the SSI camera office is occasionally referred to as Pat's Porch for its main resident, Pat Woida.

"These images are taking forever to get here," Peter says.

"I think the scoop is empty," Boynton says.

"Where did it go? Mysterious," Peter asks rhetorically. He smiles.

"SHIT!" SOMEONE YELLS FROM THE SSI CAMERA OFFICE. THE FIRST image shows a small pile of dirt sitting on top of TEGA, not inside it. Nothing went in. It's all there and accounted for, but for some reason it just wadded up all shy-like on top of TEGA. It's not Matt Robinson or the robot arm's fault. He put the dirt where he was supposed to. And it's not TEGA's fault, they were ready to receive the handoff.

"We dumped on TEGA and it doesn't go in the oven. Hunh . . ." Peter says and shrugs.

It's Mars's fault.

"I'M NOT WORRIED," SAYS PAT WOIDA CONVINCINGLY, RUSHING THROUGH the crowd in his Hawaiian shirt and cargo shorts. I struggle to keep up with him. He is sure that TEGA got the sample.

"Look in the scoop," he says. There's no fuss. "This picture was taken after the delivery but it's full of fines." The image is of an empty scoop with some dirt in the bottom. "TEGA is designed to accept material just like that!"

"Look at those particles," he continues. "Isn't it obvious?" I appreciate that Pat assumes a level of understanding commensurate with my security badge. It makes me feel like one of the team. Although, if I don't say anything, I'm not going to know what's going on.

"It's not obvious," I say outing myself as an ignoramus in matters of Martian geology. Pat does not ridicule. He graciously offers a lesson in Martian soil mechanics:

The first lesson: it's not soil. You probably shouldn't call it "earth" either. Things would get confusing, fast. The official term is "regolith." Pro tip: if you want to feel like you really belong in Mission Control just drop "regolith" into casual conversation. It's instant SOC cred. Turns out the word "soil" implies that you have microbes and other living stuff in there. So it's inaccurate to call it soil until we can show that there are microbes living on Mars. There might be; but we don't know that. Yet.

Anyway, this *regolith* is a mélange of dust and rocks and other kinds of particles. A big part of Phoenix's directive is to use the tools on board to compile a basic understanding of this stuff. If we can identify what kinds of minerals make up the dirt clump resting on TEGA, we can infer all kinds of information about the history of mountains, rivers, or even oceans on this site.

"Now, you got a couple kinds of particles to know," Pat says. He seems to enjoy breaking it down. "The 'fines' describe the smallest particles in a size classification that goes from clays, to silts, sands, cobbles, and then to boulders. For geologists, it's important to know how much energy is required to move these particles around if you're going to be scooping them up and dumping them into your instruments."

The amount of energy required to move particles is very high when they are large like pushing a boulder. As they get smaller it gets easier; like blowing a grain of sand. Yet, once the particles get sufficiently small, the clay stage, it gets more difficult again.

"It's hard to scatter clays; you couldn't blow it off your hand like sand," Pat says. If these are densely packed clays, this could be the cause of the clumping. Or just one of many causes.

It's perilous to make any conclusions with so little evidence. And there are other possible explanations. There could be some static charge holding them together. Frost could freeze the particles together. All they know is that they've dumped the regolith onto TEGA but it's just clumped on top.

"Don't worry. It'll get in," Pat says confidently. "There were plenty of 'fines' to get in there, so there's nothing to worry about." But if Pat is wrong and it's more "cloddy" than they imagined, they'll have to try something else.

After my lesson, Pat tells me that he's not working today. It's his day off. Could have fooled me.

"I'm just here for a few minutes," he says.

Since we're the only folks on Earth working the Mars night shift, there's not much to do when we're out of phase with Earth. The only people you can interact with are colleagues. Pat says he'll stick around for a bit and then he's got eight hours to kill. "I might just watch all three Lord of the Rings—director's cut. That should fill up the day."

A PROCESSION OF TEGA AND RA ENGINEERS MARCHES OFF TO THE small conference room just behind downlink. The room is known as the "Penalty Box" and sometimes called the "Wood Shed." It's where you go when your group runs afoul of space law. The idea is that a small group can better address the problem (or assign blame). A new plan must be finalized soon, so time is crucial. They will need to quickly come up with a new plan of action, but it's not clear yet what exactly the problem is. The mission rests on the ability to pick up dirt and deliver it to scientific instruments for analysis. So far, the images tell a confusing story of failure involving some strange property of Mars dirt. This is far more worrisome for the fate of Phoenix than just a missed delivery, programming error, or "Mars-lagged engineer." They want to try to get the dirt in right away, but they need to make their best guess at what's going on with the clump and how best to fix it.

WHEN THEY EMERGE FROM THE PENALTY BOX, RAY ARVIDSON CALLS for the midpoint meeting to begin.

"There is no first sample," Ray tells the group, lingering for a moment so they can all have a moment with their disappointment. Chris Shinohara, who is the instrument manager for the SSI (and the General Manager of the SOC and Peter Smith's consiglieri) shakes his head.

"This is really the first big disappointment of the mission," I say to him, trying to sound insightful.

"No. When the filament went on TEGA, that was a disappointment. Then when the TEGA door wouldn't open, that was a disappointment. And when we couldn't control the OEM, that was a disappointment too," Chris responds. I'm in over my head.

Joel puts himself in charge of the problem. He is the "anomaly lead." When things go wrong in Mission Control, they're called anomalies. It's an appropriate euphemism because nothing really goes "wrong." Things just don't go as expected. It's a good lesson. The anomaly lead official determines what happened, tells everyone how to keep it from happening again, and then, most importantly, files an incident report on it. Everyone takes his place at the big conference table.

"We went back to the woodshed and now we have two hypotheses," Joel says. "The first is that the material is sticky. The second is that the solenoid has a mechanical problem." The solenoid is a shaker that sluices the material through a grate that sits on top of the TEGA oven. "In light of these hypotheses, we made a decision: we will stand down," Joel says. That means do nothing. Everyone agrees they need time to understand how best to break up the dirt pile and shake it down into TEGA, where it will be heated and properly analyzed. They want to take some time in the lab on Earth to figure out how that might work best. And with a stroke of the delete key, the old plan is tossed aside. There will be no TEGA bake experiment on Mars today. Tonight, it's all hands on deck figuring out what to do next. It's a little pause to give us our best chance for a bake in the next few sols. The meeting is dismissed and everyone scurries off to his computer. Work on a new plan begins.

CHAPTER THREE
CONTROL ROOM

SOL 13

THE SOC IS NOT THE DECK OF THE STARSHIP ENTERPRISE. THERE are no titanium space consoles or glowing red orbs. There are control rooms at Kennedy and Johnson Space Centers that look like the mission you're imagining in your head—these are the first class, flagship missions from the good old days of the Cold War, when nuclear destruction was imminent and we were competing for space supremacy with our comrades in Russia. We don't have one of those control rooms. There's a fancy control room with blinking lights and fancy gadgetry at the Jet Propulsion Lab. We don't have one of those either. This is a barebones budget mission. But hey, we've got spirit.

I like to think the church basement aesthetic for a scientific laboratory is not a budget constraint but a brilliant gambit from our captain, Peter Smith. Although, Peter might disagree. It's a bold move indeed. By rejecting the clichéd Mission Control look, Peter limited the temptation to rely on outdated mission memes for our modality of discovery. This is not Battlestar Galactica. We are not Viking I. We are not Viking II or Pathfinder or even the Polar Lander. We are Phoenix. We're using light brown industrial carpet, regular old cubicles, and aluminum or plastic folding chairs. There are no plants, and some of the acoustic tiles on the ceiling are broken. With this particular design motif, the central feature of the downlink room could never be a command module for the captain. There isn't even a lectern. Instead, we have a few folding tables of discovery pushed together. Sprouting from the center of these tables are green and yellow power cables attached to the projectors and A/V system.

At first you might feel there's a sad kind of impermanence to the whole place—like 90 days from now when Phoenix dies, a van is going to pull up to the SOC and repossess everyone's space dream. That feeling is fleeting. Soon you fall in love with the minimalism. We don't need any of that NASA junk. This DIY version is a far more hopeful space. It reminds us that no idea is so big that it can't be made operational with furniture bought at Staples.

At the head of the table is one large red director's chair. It's much higher than the 18 other folding chairs. It says "Smith P.I." on the back, fitting for the captain of this vessel. The letters are embroidered in the Star Trek font, called *galaxy*. That's not NASA standard issue.

"It was a gift," Peter says, for an indefatigable effort to get his colleagues to Mars.

At the foot of the table, next to the projector screen, is a seat reserved for one of the cruelest gigs in the space biz: Science Plan Integrator I (SPI I). It's the SPI I's job to manage the lander's schedule. Try being the executive assistant to a famous spaceship: non-stop requests for experiments, appearances, and interviews.

As SPI I you must have a pretty firm grasp on lander subsystems, resource constraints, and all kinds of other esoteric space planning minutiae. There is a whole slew of flight rules that, if violated, will bring the wrath and fury of a team of engineers lying in wait. There are a lot of systems and subsystems. Each comes with a whole team of engineers. These folks stand around the SPI and ask for their activities to be included in each day's plan. The SPI must scream at them to back off and give her some damn space. You can't do science with all these people up in your grill.

Today's SPI I is Suzanne Young, a chemist from Tufts. She has very long brown hair and is partial to flowing dresses. She also has a strict rule about more than one engineer within a two-foot radius while she takes requests for lander activities. I imagine that she's one of those tough-as-nails young professors who fails a lot of students but inspires a few to become the Peter Smiths of the next generation. Some of those brilliant students are here working on the mission.

Suzanne spent yesterday, her day off, here working. She says her days off are the only real time she can get some science done. Suzanne helped develop MECA's wet chemistry experiments. Like many of the science team members, Young acts as a systems engineer. She pretty much volunteers for any job they'll give her—eliciting an occasional deep exhale from senior team members who'd rather not have her bite off more than she can chew.

She attended almost all training missions and never shied away from double duty in spite of the long hours and stress. Many of the scientists who use research grants to fund their work here have the the the same impossible workload as Young. It makes this low-budget mission possible. Normally these jobs would be staffed by JPL engineers. Phoenix didn't have enough money for all the positions it deemed necessary. They had a choice: operate short-staffed, or ask the scientists to step in and train for engineering gigs.

"There aren't all that many chances to operate a spacecraft on Mars," Suzanne says. "So it's kind of a no-brainer."

Each sol is an elaborate color-coded array of activities: digs, photos, measurements, data uploads, downloads, maintenance, and upgrades. In case you find yourself in Mission Control, it might be helpful to know that TEGA is purple; MECA is green; SSI is red; the RA is yellow and blue for MET. The SPI I arranges the activities in an efficient manner and in line with the greater goals of the mission. The SPI I is part Sotheby's auctioneer, part horse-trader.

Seated opposite the SPI I is the science lead. The sci-lead, as he's referred to on the job, oversees the process from the scientific perspective. The sci-lead has an engineering counterpart, the shift lead. Together they work with the SPI I to incorporate the requests for various lander activities. They must be mindful of short-term constraints like how much power the solar panels collect, the state of the battery, equitable distribution of lander resources, and long-term goals. Every few days, sci-leads, SPI I's, and all the other support positions swap roles or rotate out of the lineup.

"You have to keep people fresh and engaged," Joel says. "Plus, it's safer to have multiple people that can do each job." This will be especially true at the end of the mission as the sol schedule takes its maximum toll on everyone's physical and mental strength.

The 16 other folding chairs around the conference table represent the various instrument and science theme groups. There are four theme groups: Geology (GSTG), Chemistry (CSTG), Atmospherics (ASTG), and Biological Potential (BSTG). In the Phoenix organizational scheme, the theme groups consult with the instrument groups to create strategic plans. Then those plans are all compiled to create a tactical plan. If all goes well, that's turned into robot code and then radiated out into space for Phoenix to receive at its next scheduled communications pass. All this planning confusion starts to make sense after a few months of observation. At first, it's a little overwhelming.

In the depths of the SOC, down one hallway and then another, then a left when you hit a dead end, is the Payload Interoperability Testbed (PIT). The PIT is a warehouse-like space that holds the lander's gimpy non-flying twin; I call her PHX II. Bathed in spotlights and the over-chilled air of the PIT, when you catch sight of her at just the right angle, Phoenix II has the uncanny ability to give you the we-work-in-space chills. It's the best antidote for the inscrutable planning sessions.

The PIT offers the most cinematic, sci-fi feel in the building. Interns from the local art museum even arranged the rocks on Phoenix II's platform to match what we see in the pictures downloaded from Phoenix I. It really freaks out the conspiracy theorists.

Phoenix II lives on a platform three or four feet high. A sign overhead reads: "FLIGHT HARDWARE: PROTECT IT!" It reminds engineers of the gravity of this space. Aluminum police barriers protect the stage from any screaming fans who might try to rush up or throw their underpants. The education and outreach team offers tours most Wednesdays. As soon as our schedule swings around to daytime, we'll be sure to attend. If you're lucky, you'll see PIT test engineers wearing Phoenix lab coats and grounding straps approach the lander to prod some gizmo or blinking light with a probe. The PIT is used to test all the digging and shaking before the commands are compiled into robot instructions to send out into space. If you get really lucky, you might just see Bob Bonitz, the world's foremost robot arm man, digging in concrete trays or the various sand pits. If you're extra nice, he'll let you carry a tray.

This is one of a few extra super-restricted spaces in the SOC. There's a short list of people with access to the glass-enclosed work area and lander stage. I'm not on it. Yet.

In the kitchen, just down the hallway from the PIT, there is stale coffee available for a quarter and an honor bar. I pour myself a cup of coffee and open one of the three freezers, hoping to find the free ice cream. There is no free ice cream here. Mostly frozen burritos. Someone posted a theme group sign on the freezer, "Biological Potential." Mars humor at its finest.

"Want a sandwich?" Morten Madsen asks. The Danish team makes sandwiches in the kitchen. It may be the middle of the night in Tucson, but it's lunchtime on Mars. The bologna sandwich break is a good opportunity for small talk. Morten Madsen is a professor at the University of Copenhagen. He is the leader of the magnet team but not the sandwich making operation. Morten's team built a magnet experiment—part of the MECA package. His former and current students are also involved in the RAC and TEGA, and they built the telltale for the MET. Madsen's lab has a wide breadth of space projects.

"Running a research group is like owning a small business. You worry a lot about getting enough money to pay all your employees and keep your doors open," Morten says. "But we get by."

Morten Madsen is the undisputed master of Mars magnets. His mentor, Jens Martin Knudsen, passed the torch when he left the pole position. Now he carries the mantle for an innovative and expanding Danish space program.

"Most of my career is about iron," Morten says. As a graduate student, he studied how beavers use iron in their teeth. Although it's not exactly clear how, this led to an interest in studying how iron in insects enabled them to navigate. He planned a field study in Australia to do the work.

"My funding fell apart and I couldn't go. So I found something even farther than Australia that had a lot of iron," he says. Morten decided he'd investigate the iron deposits on Mars.

That was his calling: iron and magnetism on Mars.

"Yes, I would say magnets have a strong pull on us," Morten says, preempting my opportunity to make a bad joke. The draw is that iron deposits could tell us something about the history of water on Mars. Morten and one of his graduate students, Kristoffer Leer, explain how this rather odd-sounding phenomenon works. Line Drube, another one of Morten's students, organizes their lunch station.

"Different minerals form based on their iron content's exposure to water," Leer says. "If you can identify the minerals in the dust, you can start to understand the history of water on Mars. It's like a look back in time."

Why is this important? Understanding the history of water on Mars is one of Phoenix's "Level One" requirements. Phoenix has 10,000 requirements. These are basically the promises made to NASA about what Phoenix will do. NASA is also fond of well-defined success metrics. The 10,000 requirements of Phoenix are broken down into five levels. "Level One" requirements concern the big picture, and include things like "understand the history of water on Mars." "Level Five" requirements are more specific, such as how an instrument might perform, say, or the heating capacity of the TEGA oven.

Morten's experiments are part of this big picture, and they came relatively cheap. Magnets don't need power to make them work, just a camera to take pictures of the particles they attract.

"For us, low-budget is a great budget," Morten says. He's used to scrambling for money to pay for his science projects.

"We're always on the verge of closing shop," he says. These days funds for magnets are hard to come by. "If you want to get on a mission, just offer a lot of value for little money." It's useful advice from Madsen. There's a Mars mission adage that goes "Danish magnets are like Danish cookies, they're just better." Probably because they come with Morten and his team.

Morten got to work with his mentor, Jens, devising magnetic experiments that might be able to accompany space missions. Their inspiration came from an unusual place. A few senior managers at the Jet Propulsion Lab heard Morton's mentor give a lecture about understanding iron deposits on Mars. This was just after the collapse of communism in Russia, and the JPL team was looking to engage Russian space scientists to keep them from working on nukes for despotic regimes. They thought it was a good fit. A few months later, Morten built the prototype for a Martian magnet experiment in his basement.

That project didn't get off the ground, but Morten's magnets lived on. JPL asked Morten and his cost-effective team to talk with Peter Smith about partnering for the Pathfinder imager.

"We hit it off, and we've been friends and done work together since then," Morten says.

The Danes combat Mars lag by living like a family. They maintain a strict regimen of meals to keep each other on track. Their neighbors were a little taken aback when they woke to find seven Danes drinking beer and barbecuing steaks at dawn. (Due to the extra 39 minutes added each day from the Martian sols, shift I now ends early in the morning Tucson time, and dinner time is now at daybreak.) And even though they have an Icelander and a couple Germans on their team, Morten's group is affectionately known as "the Danes."

While the cold-cuts are assembled into lunch, there's talk about the magnet properties of dust on Mars and how tiny dust particles would be a big danger for astronauts on Mars. These particles, I'll have you know, could be inhaled, damaging lung tissue, or work themselves into the first Martian astronaut's machinery, fouling up critical parts.

"Seriously," Kristoffer says, "part of the Phoenix requirements are to assess some of the dangers posed by the Martian environment to future astronauts."

BACK IN DOWNLINK—THE CENTRAL ROOM WHERE ALL THE DATA IS downlinked—a detailed analysis of Mars's most perplexing dirt clod endures. The clod talk flows from kickoff to midpoint, then lasts through the daily science meeting, then long-term planning and short-term planning.

Both the science and the day-to-day administrative tasks are tough on a new guy. (Me, for example.) There are nano-moments during the day when I think I know what's going on here; I should propose some kind of solution to getting the mission back on track. Mostly, I'm wrestling with cluelessness. I have passing thoughts that I could go to school to become a Mars soil physicist. Then I'd know what was going on and how to solve the problem of sticky dirt. I even come up with a title for my thesis, "On the Cloddy Nature of the Fines in Martian Regolith." The work nicely combines all the words from Pat's lesson into a very compelling title. If only I could get the science team to read it, then we'd surely be out of this jam.

At 8 hours and 7 cups of stale coffee, my brain is fogged-in and over-caffeinated; my eyes are glazed. (Mmmm, doughnuts . . .) I have thoughts about canceling the whole mission and returning to Earth time. There must be a big red abort button somewhere around here. Though you'd think I would have seen it in one of my many walkabouts. This Mars tagalong is tired and a little loopy. After another cup of coffee, I feel once again inspired—coffee makes me manic—and vow solidarity with the team. *I'm going to stick with this dirt summit until the bitter end.* These are my people now. I'm here until the last member of the shift I team calls it a day.

But man, these eyelids are heavy. At 10 hours and 10 cups, I decide to quit.

The tiny napping chambers tucked in the uplink room beckon. I peek inside but decide I can't in good conscience take a bed from someone who might actually be critical to the mission. Best to sleep it off at home and return with a clear head tomorrow. It's early evening on Mars. But the sun is coming up in Tucson. I duck out, leaving my fellow spacemen on day three.

WHEN I LEAVE THE SOC AND VENTURE OUT INTO THE EARTH WORLD, it's a little disorienting. It's hot and bright. My body wants it to be the middle of the night. I walk into town, desperate for something to cover the windows and get an AC for the sleeping chamber (bedroom, to Earth folk). Everything is closed. It's too early in the Earth morning. This is a good lesson about using a Mars watch to schedule Earth-related shopping activities. So I give up and head to a cafe for dinner.

The tattooed barista smiles and waits patiently for me to order. She fails to notice or ask me about the Phoenix or NASA stickers prominently displayed on my notebook. That's okay. I just casually mention my affiliation for her mental stimulation and perhaps a 10% space discount. She's not interested in space.

"It's breakfast and the kitchen's closed," she says with a smile.

"Give me a cup of coffee and a couple scones," I say.

I look around at all the folks drinking their coffee and stuck here on Earth. They seem so far away. I don't speak to any of them.

CHAPTER FOUR

CLODDY WITH A CHANCE OF SPRINKLES

SOL 14

BILL BOYNTON, HEATHER ENOS, DOUG MING AND TODAY'S mission manager, Dave Spencer, are huddled up at the back of the downlink room.

"The blocks all worked. Still the oven is reading empty. Well, we've never had an issue like this. All the material slid down the slope in all of our tests. Maybe the solenoid is not working?" Bill Boynton says. Bill suggests it's possible Phoenix is running the commands and the solenoid that shakes the screen just isn't functional. The mission manager folds his arms across his chest. He looks disappointed. Mars isn't behaving.

"I guess we should have a TEGA caucus," Heather says. Bill agrees.

"You're not invited," Heather says to me with a smile and a wink. They walk away.

In a back room, Bill, Heather, the TEGA engineers, and mission managers discuss options with the spacecraft team. A science and engineering task force works on a solution. They're going to take things slow. It's too soon to panic, but now is certainly a good time to worry.

The mission has two tasks now. First, and most important, is figure out how to de-lump the clumps before they try to put them inside their instruments. The RA team is devising new methods to deliver samples to solve this problem.

"It was a long night for some of us," one of the RA engineers says. All the activities needed to get coded and validated. Whenever there's a first-time activity, there's an extra layer of scrutiny added to protect the spacecraft. When it comes to robots, Do No Harm is engineering's mantra. The test is code-named "the sprinkle." Today the RA team will unveil their sprinkle plans to the team. My sources say the sprinkle is where it's at. They have no doubt we'll sprinkle our way to glory. And I have faith they know what they're doing, even if I do not.

The second task is to get the pile on top of TEGA into the TA-4 oven. This problem is less severe, but delivering the dirt would feel like progress. While morale may seem superfluous, the scientists are only human. These moments are important to keep the team sharp as our adrenaline reservoirs run dry with each passing sol. It wouldn't hurt the PR effort either.

An impromptu SOC-wide brainstorm happens. Everyone has an idea for how to improve delivery or get the dirt down the funnel. Mike Mellon suggests putting a rock in the scoop and shaking it, like a marble in a paint can, before the next delivery. That way they'll break up the clumps before they dump. Good idea. I hear someone say we should chop up the dirt with the scoop blade. This technique comes from a similar pulverizing process that involves a light-toned narcotic powder and a credit card or scoop blade if you have one. It was very popular in the 1980s. A TEGA engineer suggests they fling rocks on top of the grate. It'll tamp down the pile. When he sees me take a note, he looks worried.

"I was just joking," the TEGA engineer says and quickly retracts his statement. He pulls me aside and says he doesn't want some nonsense like that ending up in the newspaper.

"Newspaper? Do I look like a newspaper reporter?" I ask if a book is okay. He rolls his eyes. I take that as a tacit yes.

IN THE RA OFFICE, ASHITEY TREBI-OLLENNU AND MATT ROBINSON TYPE away. I take a seat, but they don't really acknowledge my presence. "So, you guys working on the sprinkle test?" Yes. Yes, they are.

"Wait, why don't you have a media badge?" Matt wants to know. Even though we've engaged in casual conversation for several days, they have a flash of lucidity. They don't really know who am I or why I keep trying to pump them for information. I ask myself these same two questions several times a day. Who am I? What do I want from these people?

The Jet Propulsion Lab and NASA require that journalists be escorted inside their facilities. So someone who asks questions but isn't an engineer or scientist is an alien. There are strict rules for NASA and JPL PR.

"Sometimes we break them, but we usually only find out after the fact," Carla Bitter from the outreach team told me before the mission. "There's no 'working for NASA 101' and they can make it tough if you're trying to do something creative that inspires people." For the most part the engineering team doesn't get too involved in the mission message.

"No, what are you *really* doing here?" Matt asks. I tell them I'm here to write about the people who work in Mission Control.

"That doesn't sound very interesting," Ashitey says. There's no hint of irony in his voice; a purely professional assessment.

"Maybe a movie is a better idea," Robinson adds. He thinks for a moment. "Maybe Bob Bonitz [their boss] should be Clint Eastwood," Robinson says, content with his casting choice. Bonitz is a wiry man of few words, considered one of the top robotics engineers in the country, and counts 15,000 sky-dives (included many world record jumps) to his credit.

"And you know Ashitey is royalty in Ghana," Matt Robinson says. "And that *Coming to America* is the story of his life. If he's nice, we'll let Sidney Poitier play him. Although Gary Coleman might be a better choice," he says, carefully considering his options. In spite of his boyish looks, Ashitey is a JPL veteran. He's got a lot of Mars experience, not enough to make Sidney Poitier a good casting choice, but still he's a pro. He was even a driver for the last rover mission.

Robinson is affable and earnest—a good-natured guy who married his high-school sweetheart.

"You know his wife still brings him his lunch," Ashitey says. One reason Matt is not too fond of spending the rest of his summer here in

Tucson Mission Control. Bologna sandwiches and ice cream are good but haven't got much on a home-cooked meal.

Phoenix is Matt Robinson's first space job. Since he finished graduate school—who knew you could get a Ph.D. in the claw game?—the robot arm is all he ever does. When he says he spent the last five years scraping concrete, that's not a joke. The permafrost Phoenix is after is as hard as or harder than concrete; so they practice in concrete.

"We don't want to just turn the scoop over and dump. If we just keep dumping for every delivery, we'll eventually start to cross-contaminate the other instruments on the deck," says Ashitey. He clacks away, working the vector angles and timing for the dirt sprinkle. He's used to reporters getting the story all wrong and takes his time to explain. He's patient with me but keeps his expectations low.

"If there are no precise controls, you can't easily deliver a small sample." Ashitey takes out a plastic engineering model to show me. "You want a scoop? I'll give you *the* scoop," he says and laughs.

First, it's not a scoop. Officially, it's the ISAD, Icy Soil Acquisition Device. This is a precisely engineered piece of machinery, not a shovel you'd use to clear snow from your driveway. The angles, width, height, and shape, every aspect of the mechanics carefully crafted to capture and guard the Martian muck.

"There's a lot to think about when you design an instrument," Ashitey says. The scoop can't be reflective, lest the reflection of the sun blind the camera or generate heat and melt the contents. With so many variables in the design process, you'd need a whole edition just to describe scoop and RA genesis. Even the simple stuff, like buying the right ball bearings, isn't as easy as it sounds. You know the shiny number five ball bearings that really trick out your gear box? They were impossible to come by. Phoenix battled government procurers snapping them up for the Iraq war effort. Once they got the goods, there were new challenges. The materials and construction processes were crucial. The methods they chose absolutely couldn't interfere with the science data. Any stray chemicals used in the manufacture would foul up the Martian signals. So each part must be closely scrutinized, its materials and possible defects well understood, every last potential variable accounted for.

The end product is a fine scoop. Some of the design highlights include a titanium scraping blade and a rasping tool—a small drill-bit-like device—for excavating the cemented ice-soil. The ISAD's rasping tool and load plate should deliver the ice shavings of soil that's been frozen for eons at hundreds of degrees below freezing. And do it with aplomb. Later in the mission, the scheme goes like this: plant the load plate onto the ice and drill small shavings that will ride on the rasp bit into a channel in the back of the scoop. From there, a series of precisely-machined little grottos at the back of the scoop allow the RA to transfer the ice into the main scoop compartment for delivery.

Another groove down the center of the main scoop compartment works in concert with the rasp to achieve the sprinkle delivery. It's kind of a clever trick. They turn on the small drill but they don't use it for drilling. They use it to get the sprinkle effect.

"If you put the scoop head at the proper angle and turn the rasp motor on, the particles will dance down the center groove and file out like a stream of water," Ashitey says. "It's beautiful."

This is the sprinkle.

The sprinkle was intended to deliver small amounts of dirt to the microscopes on Phoenix. Now it's being adapted to delivering sticky dirt.

Ashitey hands over the scoop. It's like a pacifier for nosy guests. They're able to get back to work while this guest plays with their toy. Awesome.

MIKE MELLON AND PETER SMITH ARE AT MELLON'S DESK WITH A COUPLE other instrument co-investigators. They focus on the clump: calculating and figuring. One of the JPL press attachés, Guy Webster, stands close by. He is responsible for molding the clump into a media-friendly story. When the dirt didn't budge the first time, the press office put out a story: "Phoenix Checking Soil Properties." They had to say something. They sorta jumped the gun the day before by preparing everyone for a big delivery but not preparing for the complexity of it all.

I would have gone with "130 Brilliant Scientists Fail to Get Dirt in Cup." The press release is a non-story designed to hedge. It's written more for NASA insiders who already know what's going on. The press release describes the soil as "resting on the screen." Which makes me think of telling the kids the family pet fish is "resting," when it's really just dead in the water.

So now what? The idea that dirt on Mars is too sticky to experiment with is subtle. Is the average Mars news consumer going to buy the "resting" approach?

Peter and Mike discuss the resting clump. I'm poised to sidle up to the captain, but I then I take pause. Even though eavesdropping is how I glean most of my scientific insights, I realize it's day four of my five-day trial, and I don't want to be in Peter's face when things start to go pear-shaped. What if, in some strange transference, Peter decides I'm to blame for all these issues? *Kessler, you're a real goocher, we're gonna have to ask you to leave,* I imagine Peter would say. Then I'd surely get the boot. And if I get the boot, I'll never get the wildly compelling insider's perspective on what makes men send rockets to Mars. There's only one thing to do: stand behind Guy Webster. It's not noble to use poor Guy Webster to block myself from Peter's line of sight and potential ire, but at least I'll live to report another day.

Peter examines the pile on top of TEGA. There must be some hidden clue in these images. The dirt is not frozen; it's just somehow too sticky to get inside TEGA. For all the shaking, it's not really budging. Mellon clicks through the time-lapse photos of the shaking sequence. He goes through again. And again. Over and over. Peter stands up abruptly and marches to his office. I avoid eye contact by looking down and writing furiously in my notebook. Peter returns with a measuring tape.

"It's come to this," Peter says with authority. Uh-oh, he's blown a fuse. I hope he's not going to smash anything or take it out on Mike Mellon because the dirt is sticky.

Peter pulls out the tape—he doesn't use it to beat Mike into submission—instead he gently holds it up to the computer monitor. Peter compares the movement of dirt in the images. Measures the changes on the screen. Guy Webster looks worried too.

"I can't tell which is the 'before' and which is the 'after,'" Guy says.

"It looks like they're not going to figure this out soon," he continues. NASA wants some follow-up for the "resting dirt" story. There's not much more to say.

"WE'RE WORKING WITH A DUST PAN AT THE END OF A FISHING POLE,"
Ashitey says. Most of the science team doesn't really understand the
limitation of the arm. Ashitey and Matt tell me that operating a robot
arm on Mars is not what you might think.

"It's really hard," Matt says. "It's not obvious and it's a point that's
lost on many from the science team. They [the scientists] think the
samples deliver themselves." Really? He assures me that in fact there's
a disconnect. The problem is conceptual.

"Digging on Mars is easy to understand. And that's the problem,"
Matt adds. Everyone thinks it's as easy to do as it is to talk about.

Ashitey says no matter how many times you tell someone how TEGA
computes the mass of elemental particles, there's going to be a tiny bit of
magic you have to accept. Or maybe a lot of magic. You can't see TEGA's
atom-weighing scale. You don't put some atoms on the scale and then
slap a tag on it. It's not a butcher shop. On some level you just have to
accept that TEGA can weigh atoms. On the other hand, we all know how
to dig in the sand with a pail and shovel. Understanding how a scoop
and arm pick up dirt is easy. Any four-year-old can do it, right?

"This arm is a legacy," Matt says. The robot arm came from the
design of the 2001 lander mission but had to be significantly modified. They needed new gear motors and lots of testing to be sure it
could dig in the cement-hard ice. "There are no pressure sensors and
the gearing is not as precise as it could be," Ashitey says. That means
it doesn't have the ideal features.

"They bought us a Ford Focus but we are asked to perform like we
have a Ferrari," says Ashitey. The RA was not built to the necessary
degree of precision that RA team would like. Its joints have some give
in them and that makes it hard to predict how and when they'll drift out
of alignment. This complicates any fine movements the team asks for.

"We're rated for centimeter degree accuracy but the scientists ask
us for millimeter accuracy," Ashitey says as though he's revealing a
devastating mission scandal. I try to sympathize. While the arm is
a piece of engineering equipment, it's considered part of the science
payload. It's what's known as a mission-critical instrument. Basically,
if they can't use the arm, the mission is a failure (to NASA).

Operating the arm to such an exacting degree, essentially coaxing
extra-precise movements out of the machine, is only possible because

of the RA team's sense of duty and dedication. It's a matter of crafts-man's pride. It's a craft you get good at after years of working mind-numbingly long days. They know every quirk and slip to that arm's joints and somehow seem to anticipate problems from 200 million miles away. The ISAD is an extension of their own body. Still, no gold stars get tacked on the kitchen fridge for these heroics. Pride in their ability to deliver on, without question, whatever request the science team asks is all the reward they need.

Matt Robinson studies an array of numbers. The figures cover two 30-inch screens.

"This is the Matrix," Ashitey says. Then he laughs. It's a millisecond-by-millisecond account of every digging vector, current reading and some 45 other details of their work. There are 45 columns of data in the "Matrix." Somehow he is reading it. Somehow it helps him do his work.

"We can recreate what happened on Mars with a little movie," Ashitey says as he loads the array into the computer. We watch what the arm did on Mars in the cartoon version. They use these movies to understand how the scoop operates in the Mars environment. Our geeking out on the matrix is suddenly interrupted by the boss.

Bob Bonitz comes in. He looks tired and grimaces. Probably because he was awake all night. He asks me to leave. Well, it's more a nod and a point. I get the idea. Bob is not much for small talk. Even if he were, there's too much on the line for any niceties.

"Please, close the door behind you," Bob says. I think that means he likes me. The RA team still has a long night ahead, and I've overstayed my welcome.

THE PHONE AT THE CONFERENCE TABLE BEEPS. "DENVER ONLINE," SAYS the caller. It's the spacecraft team from Lockheed Martin. Outside of the SOC, a large support contingent of specialists works to keep things moving as smoothly as possible inside the SOC. These are all the systems experts who know what to do if the batteries get too warm or the TECP needles are out of whack.

Doug Ming calls the group together for the midpoint meeting. The conference table fills quickly. It's standing room only.

"Welcome to our shift II friends," Doug says. The second shift team is just getting into work. It is shift II's job to turn the plan created by the science team into solid robot instructions. That means coordinating, coding, compiling, and testing custom software that's thousands of instructions long and bug-free, every day and in less than 14 hours.

"It's going to be another exciting day on Mars," says Doug. That's a coded message meaning the old plan was chucked out the window and we'll be starting from scratch today.

"There are going to be a lot of decisions to make and we have a short timeline," Doug adds, tugging on his belt buckle. Doug is a geo-chemist who works in the Astromaterials Research and Exploration Science Office at the Johnson Space Center in Houston. He speaks with a Houston-space patois that's a mix of down-home familiarity and obscure NASA acronyms.

The engineer who tracks our data download, the TDL, is once again Jim Chase. He prepares a document on what we got and what's missing.

"The passes looked good and we got our data," Chase says in summary.

"Looks like we have a healthy spacecraft and com passes. Let's hear about our instruments," Doug says.

"It worked. The dirt moved," says Dave Hamara. He is the TEGA engineer designated to speak for the group at midpoint; officially his job title is instrument downlink engineer (IDE). Dave has his longish red hair pulled back in a ponytail and under a hat that looks like it's been his favorite for decades. He's always wearing sandals with socks and he's got some gadgets on his belt.

"Geez, I've been working on TEGA forever," he says on a smoke break outside the SOC. The first incarnation of TEGA was built more than 15 years ago for the doomed Polar Lander mission. The images Dave speaks to at midpoint are the same ones that inspired Peter to measure his computer screen. TEGA engineers collapsed the pile by using the solenoid to shake for 20 additional minutes. In all the testing they only needed 10-30 second runs with their magic fingers. But now that's not enough.

"The effort was not enough to get dirt in our oven," Dave says to the team. Meaning it didn't actually work after all. "I'd like to show a movie." It's a time-lapse of the dirt pile. We all watch it on the main projection screen. I've pasted the script below:

Fade in.
MARS. DAY.
A dirt pile sits on a lander.
Dirt Pile: [Sits silently]
PHX: [Shakes]
Tension builds
Dirt Pile: [Suddenly and without warning, the pile collapses on itself.]
TEGA: Yes.
Dirt Pile: No.
Peter Smith: Shit.
Andrew: Can someone tell me what's happening?
Engineer 1: Shh . . .
Aaaand . . . scene.

You can imagine where the plot goes from there. We slump with the dirt. Least dramatic sci-fi film ever.

"Looks like some exciting physical soil properties," Doug says, desperately trying to keep the excitement from getting sucked out of the room.

"What's up with the images?" Doug asks. He wants more analysis from the geology theme group.

Mike Mellon spent the morning staring at these images with Peter and half the team. Since he's the geology lead and clump expert, everyone expects some profound insight. He turns on the microphone in front of him. He leans forward and clears his throat.

"It's dirt," he says. Classic deadpan Mars humor. There's silence. "It moved and it slumped."

Doug asks if there are any hypotheses. Silence.

"Well," Doug says, "I talked it over with the dig czar and I'd like to propose that we do a practice dump on the lander deck." The experiment would go like this: Use a modified sprinkle test to dump dirt on the lander deck. This way they can see how the clumps fall and get more insight into the sticking.

There are grumbles. The suggestion causes some strife, and all sorts of arguing breaks out. Soon there's too much going on. *Please, people, one at a time. I'm still new here. Now who is arguing for what?*

Chris Shinohara comes stalking into the SOC, waving lander blueprints.

"If you start to cover the deck with dirt, we'll contaminate the calibration targets," he says, clearly annoyed. Chris is barrel-chested and has a booming voice. And Chris will not stand for dirt on his calibration targets. These small color wheels help create accurate color images in the hazy pink Martian air.

"See that?" he says, showing Mike Mellon and Mike Hecht just how close the calibration targets are to the dump site.

"I'm not sure it's that big of a deal," Mike Mellon says. "There's some risk, but if we can't determine how the dirt falls out of the scoop, we may never get a successful delivery."

"Yah, we should just go for it," Mike Hecht says. Chris doesn't like this. Dick Morris, a Senior NASA scientist and co-investigator for the geology team, defends Chris.

"There's already a lot of dust, and if they screw up we're toast," says Morris. They all move into the RA office and close the door.

At yestersol's science meeting, Ben Clark, one of the elder statesmen of the mission and the chief scientist at Lockheed Martin, pulled up some images from sol 57 of the 1976 Viking II mission. He thought there were some comparisons that might be useful in the stickiness of the soil. He thinks the science team might want to study the old Mars images for any parallels. That's a pretty impressive memory.

Doug Ming calls everyone back together.

"We will go ahead with the sprinkle test," Doug says. "Does anyone have any issues with that?"

There are lots of concerned looks, but no comments. It's 6:08 a.m. in Tucson and 8:28 p.m. on Mars. The plan for tomorrow's sol slowly takes shape.

Peter leaves the room and heads to the RA office to discuss the strategy with Bob.

CHAPTER FIVE

RED HAZE

SOL 15

MY HEAD FEELS FUZZY. THERE'S A BRAIN SENSATION happening. If you've ever had your brain removed from its casing, massaged with Ben-Gay, and then gently put back in its skull, you know what I'm talking about. The feeling involves poking, tingling, burning, cooling, light-headedness, and, most troubling, cravings for oatmeal. This is the Mars lag I've heard such good things about; the early symptoms of Martian mania. Exhilarating highs. Crushing lows. And of course the creamy mush-filled middles. It's probably best if I visit the counter-fatigue office before this morning's kickoff meeting. Better safe than sorry.

Maybe these strange feelings are linked to the end of the Peter-imposed super ultra-secret probation. This is day five of five. At some point today, I'll have to (space)man-up and talk to him. The truth is I avoided him almost entirely this week using members of the science or outreach team to shield myself from his view. I want to catch Peter at just the right moment. That moment just hasn't presented itself.

What should I think about all this avoiding? Probably that this book is going to be really boring if I hide from its main character, unless it's going to be a mystery or some existential allegory and Peter could represent a messianic prophet. That version would end with me standing on a mountaintop, shaking my fists at the heavens and screaming: WHERE IS YOUR GOD NOW? Or I could just keep asking Peter if he has time to chat and hope for the best.

For now, here's to the end of my trial period. Today, I am a fountain pen.

TWENTY MINUTES TO GO BEFORE KICKOFF. JUST ENOUGH TIME FOR A quickie with Edna and Walter from the counter-fatigue group. Just the thought of complaining about my symptoms is already bringing relief. When I arrive, Edna is packing up her things. Walter isn't around. Edna tells me she's about to leave. *NO! Why? Why is this happening to me? Is this going to be forever?* She can spare a few moments to chat. Great.

The counter-fatigue group has a small office squeezed next to the uplink room. This is where they administer their fatigue-study tests, gather results from the actigraph sleep monitors, and collect the bags of scientist urine. Bottles of urine discreetly concealed in Trader Joe's shopping bags are certainly a unique feature of Mission Control. Before I realized that Trader Joe was hiding pee, I thought it was some genius product placement. More than just slick marketing, however, the pee is important for understanding the stress levels of scientists throughout the mission.

"Have you properly prepared your sleeping area to be cool and dark?" Edna asks. *No.* I don't mention my feeble attempt to cover the windows with some discount holiday wrapping paper purchased at a thrift store. That might be one reason it feels like bramble pie is baking in my brain.

"Maybe I should buy one of those airplane eye masks," I say. Then I let out a big sigh to put a finer point of how incapable I am of taking care of myself. There's a whole barrel's worth of sleep-masks in her office. She sees through my subterfuge.

"Why don't you take an eye mask?" Edna asks me. I wonder if I'm always this needy or this is just a symptom.

"You'll feel better if you sleep. Black out the windows in your room and try to sleep on a schedule," she says. Edna offers me a few handouts about my health. A quick browse confirms my suspicion: I'm doing everything wrong, from not blacking out my windows to not scheduling my meals properly.

"Here's an eye mask. We'll be back the following week. We can talk more then," she says. She's very kind. And even though she's doing pioneering sleep research for NASA, a visit to her office feels like going to the school nurse: she can't really do anything for you, but you still feel a lot better.

THERE ARE STILL A FEW MINUTES BEFORE KICKOFF MEETING, THE START of operation sprinkle. Ashitey wants Ray Arvidson to tell him how he'll know if the sprinkle test is a success.

"We need to know the metric," Ashitey says. A good success metric—some way of objectively measuring he did his job—is engineer smack. He really does need it. How will he know if he passed? Engineers can't tolerate murky requirements.

"What exactly are they looking for?" asks Ashitey. Ray isn't sure. Any moment the results of the sprinkle will be downloaded from Mars. And Ashitey wants to know how he'll know if the science team considers it a meaningful exercise. He's not satisfied.

Joel tells me engineers become engineers because they are emotional people.

"But they seem so rational," I say, falling for his trap.

"Exactly," Joel says. "They need rules. They surround themselves with rules and precision. If they're going to succeed, they need to know how success is defined."

Bill Boynton walks into downlink. He's cheerful. He has a theory about what's gone wrong.

"It's like a lava flow," Boynton says. "No particles could get separated from the edge to let any grains fall in."

"I think it's more like a Brazilian soccer-match stampede where no one gets through the door because everyone is trying to get through the door," Ray says, disagreeing with the analogy.

I can't tell if they're really arguing if it's a lava flow or stampede, or if they're just joking. It's easy for the new guy to miss subtle Mars humor.

Either way, Bill suggests they abandon the shaking effort and start over using the sprinkle method and a new TEGA oven. He doesn't want to waste any more time. Bill is optimistic things will go smoothly for TEGA from here on out.

Kickoff starts. Doug Ming calls the group to order and reminds everyone of the importance of today's results. The teams report a healthy spacecraft and no dangerous weather systems.

"Geology, you have to look at the sprinkle test and work with the dig czar to figure out a success metric. They'll also need a formal recommendation before midpoint on how to proceed with TEGA," Doug Ming says to the GSTG group.

The communication pass should start any moment. And before there's even time to refill my coffee cup, images start arriving. There are twenty or more scientists gathered in the SSI office. Max capacity is probably six. Ray Arvidson wants in too. He barrels through.

"Make way for the sci-lead," Ray says.

There are feigned grunts and groans as Ray pushes through the crowd. I try to stand on my tippy toes to get a glimpse. All I get are backs of heads. There's a chorus of commentary and murmuring. It's mostly neutral assessments. Until an English accent says, "Yes. Ohhh, yes." The scientist must see something good, because we don't usually hear English sex noises in the SOC. There's celebration. The sprinkle test and pre-sprinkle test worked! The mission can move forward.

There's a nice spread of powdery dirt across the deck. No lumps, wads or mounds anywhere. The RA team pioneered a breakthrough method of Mars soil delivery. Seeing as it might be my last day (if Peter decides this isn't working), I feel bold. It's time to make my first scientific observation. I want to know what's it's like to be part of the discovery.

"Looks like a nice even distribution of fines," I say. I brace for the worst. No one flinches. There's even a nod of approval.

This is probably the most excited anyone has ever been over a pile of dirt since—wait, there aren't any other famous piles of dirt that are less than a centimeter high that can make this comparative work. Regardless, this is a successful sprinkle pile. Matt Robinson suggests calling it "Mount Ashitey" in honor of Ashitey's contribution to dirt science. Ashitey bristles; he does not like the idea.

"Good thing we did that pre-sprinkle," Mike Hecht congratulates himself.

"I'm not sure it did all that much," Ashitey says.

Mark Lemmon puts an end to the celebration. He kicks everyone out. He wants his SSI team to prepare the images in time for the midpoint meeting.

WITH SOMETHING TO CELEBRATE, THE MOOD INSIDE THE SOC IS buoyed. At the midpoint meeting, everyone is in good spirits. The SSI IDE shows the sprinkle test as series of photos. It's a flip-book sprinkle movie. Mike Hecht claps; he's the only one.

The RA team representative makes the formal announcement.

"There was a successful sprinkle. We are very content with this strategy. It really showed the wisdom of thought in doing this," Ashitey says. Because it went so well, they're going to make their first precision delivery tomorrow. This time MECA gets a turn—no wonder Mike Hecht wants to applaud—they're going to sprinkle a few grains of Mars dirt onto the OM (optical microscope). Microscopic images of mysterious dirt on another planet; that should count for something.

"That's a beautiful sprinkle," Doug says, taking a minute to enjoy the moment.

"Can we clap now?" Mike Hecht asks. There is applause. The mission is back on track. There is new hope for getting dirt into the instruments and having a viable mission.

SO, THIS IS THE END OF THE FIVE DAY TRIAL. I'M RELIEVED IT ENDED on a positive note. I take a deep breath and head over to Peter's office. I poke my head into the open door.

"Can I come in?"

"Yes," Peter replies, betraying no real emotion.

"So, I guess this is the end of our five-day test."

Peter doesn't really react. He shifts some papers, signaling that he's running a goddamn Mars mission, not babysitting. Point taken.

"Well, I haven't heard any complaints. No problems. So I don't see it being an issue for anyone." And with that, Peter Smith makes the historic decision to let the world's first outsider into the hallowed halls of Mission Control for a landed planetary mission.

Huge alpine mountain horns blare, the sun shines in the window. It rains flowers.

And then something really amazing happens: we just chat. And it almost feels natural. We're talking about the clump and whether they will keep shaking the dirt on TEGA. Shooting the breeze like we're two normal humans and it's going great.

This is it, I tell myself. Our big breakthrough. Peter will be spilling his guts and crying on my shoulder any minute now. Ask him about his father. No. Ask him about the nature of discovery.

Soon I'll understand what's at the heart of man's quest to conquer the solar system. Peter is not quizzing me and I'm not trying to sound like I know what I'm doing. I even tell him about an idea I have for a coffee table book, a little curatorial project where we ask everyone what their favorite images from Phoenix are, and then we turn it into a Mars coffee table book. It's a project we could do together.

"Interesting," he says.

I tell him I'll get started right away. It's going great. I've got some "dad" and "nature of discovery" questions all lined up and ready for the right moment.

Peter looks over his shoulder at his computer. He shuffles more papers. Damn, I've lost him. There's an awkward pause. It's over.

"Well, I have some things to attend to," Peter says. The NASA chiefs are waiting for his daily report.

It's probably best I go. There won't be any shoulder crying, but at least he didn't say I have to leave.

That evening, there's one last TEGA shake. And it works! Bill Boynton plays disc jokey and selects K.C. and the Sunshine Band's *(Shake, Shake, Shake) Shake Your Booty* to mark this important moment. Everyone does indeed shake their booty and Mission Control is a place for celebrating. The science phase of the mission begins.

PETER STARTED BUILDING CAMERAS AT THE LUNAR AND PLANETARY Lab here in Tucson as a research assistant thirty years ago. It was a good year for both of us. I was born; Peter's career began to take shape.

In the last thirty years he's worked himself up to freelance Mars mission captain. Ah, the good old days, when you had to take a million exposures just to get a one-megapixel image and it only cost a nickel to see the funnies.

There's not really an obvious career path that takes you from hippie at UC-Berkeley to space captain, yet somehow the winding, chutes-and-ladders path Peter took got him there. For most of his tenure at the lab, he worked under Dr. Martin Tomasko. Marty, as Peter calls him, was his mentor. At the lab, Peter plied his trade as a young space photographer, capturing images and data on the composition of our universe. He hates when you call him a space photographer or try to romanticize his cameras.

In 1989, Peter was helping Marty build a descent imager for the Cassini probe. The probe would be parachuted down to the surface of Titan, one of Saturn's moons. It was an ambitious and complicated project. Peter coordinated with researchers across the globe.

"We needed to find international partners to help us pay for the thing. With European partners we got more bang for our buck. Their universities provided funds so they could built pieces of the camera," Peter says. "Back then you'd just show up and knock on doors. There wasn't email or Powerpoint. Marty and I would pull up the overhead projector slides and make a pitch.

"That's where we met Uwe Keller," Peter says. "Uwe is a bullet-headed German. He's bigger than I am and really in your face. He had no idea we were coming. But we got along great." Keller built some detectors for the camera on the Giotto mission that captured pictures of Halley's comet. Peter and Marty could use Uwe's capabilities for Titan. And Keller held the key to getting low-cost detectors for the Titan camera. Keller said he would build the detectors.

It was an exciting time for Peter at Marty's lab, but there was the pull of something larger. Peter wasn't content.

With each new project he worked on with Marty, the feeling that he needed to strike out on his own grew stronger. At the same time, he was building the skills to make it possible to come up with his own original concepts for space cameras. Working on Cassini gave Peter confidence and international connections. All he needed was the right opportunity. The Cassini camera used a new type of sophisticated

CCD chips—just like the ones found in today's digital cameras. The instrument paper describes them as "the most sophisticated, highest resolution two-dimensional device ever carried into the outer solar system.". Peter realized he could use a CCD camera to pitch his space narrative to NASA. Pathfinder needed a camera to track a small rover called Sojourner. To get the commission, he would have to compete against some of the big, established names in space optics. He didn't have any name recognition at NASA or even a Ph.D. So what?

Peter went to see Marty.

"I have an idea for the Mars camera," he said, and Marty met him head on. "Great, let's do it."

Peter clarified, *he* had an idea, and then Marty understood. It was time for Peter to take the helm.

Peter called his camera IMP (Imager for Mars Pathfinder).

"I wanted it to have a little personality," Peter says. The camera would be the first to offer stereo resolution and create three-dimensional anaglyphs of the Martian terrain. That's right, 3D space images more than two decades before *Avatar*. Take that, James Cameron!

"Before you get too excited, the Viking I and II missions had 3D too. It wasn't quite like ours," Peter tells me. "But it could only see in 3D right in front of the lander. It wasn't on a rotating mast like Pathfinder or Phoenix. And for that matter, you should really be careful when say 'first' in the context of space. You'll almost always be wrong. Trust me," Peter warns. The IMP camera would offer a whopping ¼ of a megapixel resolution. It would feature a fancy new CCD chip; Peter would send a digital camera to Mars! And most importantly for NASA, using parts from his foreign collaborators, he could build it cheaply. The IMP looked a lot like the SSI camera on Phoenix that Peter would build twenty years later. It was just a slightly more primitive version.

The IMP would not only be a science tool but a marketing device. That's what made it great: the broader potential to tell a story. This multi-lens stereo imager was mounted at human height to offer a familiar perspective. Instead of cold scientific observations, the IMP's pictures looked like they might be tourist snapshots. (Although the scientists hate when you suggest this. The official position is that these are not *merely* snapshots.) Peter and his imaging team stitched together photo mosaics to create brilliant large-format landscapes. It was like looking out at Mars with your own eyes. It's hard to know if creating an emotional connection was part of the selection committee's decision, but Peter was nevertheless awarded a chance to send his first camera to Mars.

ON JULY 4TH, 1997, THE PATHFINDER SMASHED INTO MARS, BOUNCED around, and rolled to its final resting place nearly unscathed. Instead of expensive rockets for landing, the Pathfinder used an innovative (and cheaper) airbag design to get safely to the surface. Eric De Jong, the head of imaging at JPL, recalls the two engineers who built the Pathfinder airbags embracing wildly after landing. Both of them jumping up and down, saying over and over, "I can't believe it worked!" Pathfinder was NASA's triumphant return to Mars. It had been twenty-plus years since anyone had seen the place.

Peter and Eric De Jong suspected space fans and pretty much anyone who had ever looked up in the night sky would go crazy for the IMP pictures. Instead of offering a few images to the press and holding on to the rest for publication in some obscure journal, they decided to offer the excitement of Mars online and in real time. They made a secret backroom handshake deal that no matter how upset anyone got, they would post the images on the Internet as soon as possible, bypassing the usual JPL and NASA channels for releasing images. They were going to give the images to the people. I mean, you don't go to Berkeley in the sixties and not stick it to the man at some point in your career.

Peter and Eric played with the idea that the space narrative was already an enthralling part of our national character. We already had it woven into our imaginations, and, with the right tools, they could unleash its captivating forces for a new generation. By giving anyone with a computer access to the raw images, Peter let fans create their own Mars story.

"NASA wanted me to go first at the big press conference. We were announcing to the world that we captured the first images since Viking. They insisted we should lead with this story," Peter tells me. "I was a bit terrified. Dan Goldin [the head of NASA] was going to go after me! But they really believed in the images. I diligently wrote down what I was going to say and practiced my lines. I suggested they play Brahms' first symphony and raise a screen to reveal the images. It would be dramatic—a good story. They weren't having it. Please just talk about the images, was the message I got back. Okay, fine. Goldin looked so intense when we sat down, I got worried that I'd made a big mistake in choosing my words."

Peter takes a breath and begins his press conference reenactment. It's like he just gave it yesterday and not a decade ago.

"Imagine that you're the camera and you're loaded on the rocket ship and you're not flying first class. You're going economy. Crouched down with a solar panel on your back and parachute. And for nine months you sit like this." Peter crunches up to show me what it would look like. "And then WHAM, you hit the surface. Bam! Bam! Bam! Then you're rolling around. Phssssss. That's the sound of the airbags deflating. And you pick your head up and what do you see?" Peter pauses at that point to tell me that's when the screen was supposed to come up and reveal the images. "The talk was going great, I thought, but Goldin was sitting next to me looking at me like I was a freak. What a look he gave me! It sent shivers down my spine; I almost stopped. But if I went off my script, I wouldn't know what to say. So I just went on. I explained it like I thought my mom would want to hear if she was in the audience."

The 14,000 images that Pathfinder took came down from Mars and instantly went out to the world. The downloads literally crippled the Internet, crashing servers and becoming headline news on broadcasts and papers around the globe. One year before the birth of Google and way before the "Star Wars Kid" or even "I can haz cheezbeeger?" Peter Smith and Co. registered almost 100 million hits on NASA's site in a massive Mars picture grab.

THE IMP PHOTOS WORKED. NASA RODE A WAVE OF SUPPORT FOR THE first time in decades. The people clamored for more missions and more photos. Space was popular again. Peter spent the next few years giving talks and getting commissions for new projects. His lab was one of the biggest on campus.

Peter's successes pleased the University of Arizona. They busied themselves with transforming and expanding their space group. They hoped to one day compete with JPL for big space contracts (and, so as not to be outdone, Caltech committed to upping its profile in Playboy's party school rankings). And Peter wasn't the only one doing big things. Bill Boynton, and his colleague Alfred McEwen, were working equally hard to put the U of A on the space map. Bill's discovery of hydrogen at the top of Mars was the impetus for Peter's mission to dig up permafrost.

The Pathfinder was the first big success of a new kind of NASA: "Faster, Cheaper, Better." It was an idealistic approach leavened with a dose of economic practicality. Dan Goldin, the NASA Chief Administrator, created the program. His noble idea was to launch loads of low-cost missions and do research across the solar system. When he started, he thought missions were too costly and risk-averse. This was the nineties, and new go-go information-age companies innovated without all the bureaucracy and bloat that plagued NASA. Goldin hit the reset button to return NASA to its bold, courageous, cowboy roots. Pathfinder showed it was possible. A highly motivated team could work together to get man's robots to Mars with flying colors and great images to prove it.

The FCB (Faster, Cheaper, Better) plan was to send up loads of high-risk, low-cost missions and accept the loss of a few. The idea was that an army of landers and orbiters would cover Mars and branch out into the solar system. We would start to understand, in a systematic way, what was happening out there, one lander at a time.

For most engineers, the idea of "Faster, Cheaper, Better" is hard to accept. The slogan is a bastardized version of "Fast, cheap, or good. Pick two." It's an old cliché that engineers, designers, and admen all claim as part of their heritage. You hear it used when a client wants some miracle of innovation, in no time, and for half price. The thing about clichés is, they're often cliché for a reason, and Goldin was part of the camp that thought if innovative new tech companies are doing it, why not NASA?

The Polar Lander and Mars Climate Orbiter were two FCB projects that followed Pathfinder in just such a fashion. They went from concept to design in record time and cost a fraction of the price. There were innovations and engineering short-cuts extraordinaire to deliver the project. It was grueling, and some say miserable, work.

It was going better than everyone could have imagined. Then reality stepped in. The Mars Climate Orbiter rounded Mars and began an aerobraking maneuver (using the friction of the atmosphere to reduce its speed and properly position itself in a stable orbit). Unfortunately, it missed its mark. The Mars Climate Orbiter flew too low. Instead of skimming the Martian atmosphere to slow down, it plunged right into it. The spacecraft plummeted through the atmosphere. It crashed and burned. Well, technically it burned first and then crashed. What went wrong? One of the teams used English units instead of metric. Whoops.

Then in quick succession, the Polar Lander met a heavy fate. It crashed.

"I was sitting on a case of champagne waiting to celebrate. Then it felt like someone died, like we lost a family member," said Eric De Jong, Peter's old Pathfinder colleague. In the final moments of its life, one theory goes, the Mars Polar Lander's radar failed to identify the surface of Mars. The crash review board suggested the Polar Lander swung wildly. Its radar, which was supposed to lock on the surface and tell the landing computer the distance to impact, locked onto something other than the ground below. Polar Lander got confused and turned off its rockets a bit too soon, ending its remarkable 200-million-mile journey a few hundred feet too soon. Without the stabilizing rockets to control its descent, it only took a few seconds before Polar Lander was a sad pile of parts somewhere on the southern plains of Mars. The crash review panel identified 22 other flaws, in addition to the radar problem, that might have caused an unsavory impact. The short-cuts had caught up with everyone.

After Polar Lander and Climate Orbiter crashed, the Mars Surveyor 2001 missions that were supposed to be its successors were cancelled. The chassis of Surveyor (soon to be Phoenix), along with versions of the SSI, RAC, TEGA, and RA, was put into storage. It was probably just like the end of *Raiders of the Lost Ark*. That was the seeming end of Peter's brilliant space run. His lab of 35 became a party of one.

FIVE YEARS LATER. 6:00 A.M. TUCSON TIME. PETER STANDS IN HIS underpants at his home.

"Who the hell would be calling at this hour?" Peter said. Then he nervously listened to a caller on the other end reading him a legal statement from NASA lawyers. It was all kinds of blah blah about the fairness of the process and integrity. It could only mean one thing: they were letting him down easy. The legal speech came to an end.

"Your mission has been selected," the man finally said. Leave it to NASA to take all the excitement out of a trip to Mars. After a fierce two-year competitive process, NASA had awarded Peter his shot to be the world's first freelance Mars mission captain. The instruments were pulled from storage. (Well, most of them. NASA couldn't find the engineering model of the RA. Debate swirled around who might have swiped it as a trophy for their mantel).

After the two-year selection and scoping process to prove to NASA that Phoenix could work, things were just getting started. Next came four years of building and testing, relentless days and sleepless nights. Instruments were late, budgets overrun, marriages crumbled, new families formed, and a lot of nerves frayed. They even had to fudge a bit of the paperwork when they realized the molybdenum grease that lubricated a few of the parts would exceed its five-year lifespan before Phoenix arrived on Mars. The list of problems would fill a small library. Literally. There are about ten million pages of documents associated with the mission. These are reviews, certifications, plans, and more.

Even after years of building, NASA considered scrapping Phoenix six months before launch. NASA summoned Peter to Washington. They wanted him to defend the mission delays and cost overruns. Then he had to beg for mercy. Peter calmly defended his position and demonstrated that they would manage costs and still bring back stellar results. Somehow it worked.

A DECADE AFTER HIS FIRST TRIP TO MARS, PETER HAD GOTTEN HIS chance to go back. I watched his rocket launch from a beach near the Kennedy Space Center at Cape Canaveral. The science team was all there with bottles of champagne.

"That's the most beautiful launch I've ever seen," a veteran NASA engineer said to a colleague as they left the beach. The ground trembled and the sky lit up. Night was day. It was a clear morning and the solid rocket booster's exhaust left a red and blue trail. Everyone remarked how uncannily it resembled the Phoenix logo.

"When I saw the exhaust plume and it looked just like Phoenix, I knew everything was going to be okay. Sure, you can't convince a NASA review board with that logic, but I knew," Peter said. It was hard not to be moved.

Later that day, Peter had a boat party for friends and family. Standing at the front of a pontoon boat on the Banana River, he commandeered the microphone from the ship's official captain.

"I've been working on this every moment for the last five years, and I can say now that this is the greatest day of my life."

CHAPTER SIX

SPECIAL MARS PILL

SOL 21

SMELL THAT DESERT AIR. SUCK IT IN. SOMEHOW THE AIR IN Tucson is crisper than back east and my mind feels sharp. Sharp as a tack.

With Mars-lag creeping over my line in the sand, I make an executive decision. I take a stand. I start a course of stimulants to combat my shift worker's disorder, the dreaded scheduling disease afflicting both nurses and interplanetary workers alike. It's a great decision. By the time I arrive for work, the medicine really seems to be working. God, I feel great.

If I'm going to be here, and be alert, there's no way to sleep like a normal human being. There are only 90 days to live my space dream, and I must cover every moment of this mission. So I turn to drugs to fight the good fight. I can't miss any key moments because I need rest. The fog that's been creeping in over the last week is lifting. It's called modafinil and these pills have a fun brand name, Provigil™. Why do we call that? Because we're pros at being vigilant. *Yes, we are on the lookout, Provigil™*. Marines even take these pills before they go into combat to counteract the effects of sleepless nights in the field. Now they're our control room "GO!" pills. Since modafinil is shown to be effective in the treatment of depression, cocaine dependence, Parkinson's disease, and even schizophrenia, I'm gonna take it too!

Our version of shift worker's sleep disorder is not the friendly nurse or fireman kind. No, it is a virulent form unleashed from Hades and escorted to the SOC by Cerberus himself. I wonder if one of its symptoms is rhetorical excess; but since that has not been confirmed, I don't think the F.D.A. would approve. They might even fine you for an off-label prescription.

I learned about Provigil™ from a pamphlet the counter-fatigue group posted in the kitchen. Initially, I wanted to go in for a consultation with Edna or someone from the counter-fatigue group. But they're gone, again, for a week. So the pamphlet will have to suffice. The wall-reading in the SOC, especially in the kitchen, is a coping mechanism for all the times no one wants to talk to me. Between each of these paragraphs are hours of downtime. You probably glossed over my moments of profound loneliness and melancholy since they're tucked away in the spaces between the paragraphs. So to keep my mind occupied on Mars and not turn to thoughts of being all alone on another planet, I like to read the two bulletin boards and hang out in the kitchen. (I'm starting to wonder if these pills aren't causing mood swings.) There's loads of time-killing reading material: thank-you notes, holiday greeting cards, and helpful tips about getting by in Mission Control. And when the walls of Mission Control recommend a course of stimulants to keep me sharp and focused, who am I to disagree? So, I've started to take drugs. If it's okay with you, I'd rather not say where I got them. (Okay, shhh, I took them from a friend's medicine cabinet. He has ADHD, but I have to go to Mars. Just don't tell anyone.)

Okay, maybe the counter-fatigue group doesn't "recommend" them. That is probably overstating the case. But they do mention them. And that's enough endorsement for me in my current state. I decide to give them a go. Mostly since I'm having many of the shift work symptoms like headaches, trouble concentrating, and the occasional spontaneous hysterical paroxysm.

SUZANNE YOUNG AND A FEW OTHER SCIENTISTS SIT AT THE BIG conference table, scrubbing through the day's plan. They are engaged in an odd conversation about downloading data. Not about Mars piracy or anything like that.

"He told me we're abusing the data system," Suzanne says. It seems the JPL management is angry about the team's cavalier attitude toward data safety. Really. The JPL folks are pissed that Phoenix is taking too many risks when they download stuff and delete it. "Gentry Lee told me they had six specialists managing data on the last Mars mission. But we only have one," Suzanne tells Vicky. Gentry Lee is the head of engineering at JPL. He's been part of all the Mars missions and is not to be trifled with. One of those annoying restrictions of using a recycled lander was that memory was very expensive when Phoenix was designed. We've only got 100 megs of flash memory on board, and therefore the risk of losing data is quite large.

It's the type of conversation I'd usually walk right past. Not today. Given my expanded state of consciousness and extreme focus, I'm intrigued. *There comes a point on every Mars mission when you must ask yourself, "Is this boring or just too wide to jam through my small doors of perception?"* If I can find the beauty in moving bits across space, there's hope for me. If there's hope for me, there's hope for you. If there's hope for you, there's hope for the whole world. So I stick around and listen to them talk about data safety. Our future depends on it.

KICKOFF SOUNDS LIKE BREAKFAST AT AN OLD-AGE HOME. *WHAT? We can't hear you! You want to dig where?* There's shouting. It's a mess. I might be hallucinating. Vicky Hipkin is a physicist-turned-administrator from the Canadian Space Agency (CSA). She is the science lead today. This is her first time doing the sci-lead thing. She acts nervous. Probably because nothing works and she doesn't want to let her country down. There is a super-complicated A/V system in downlink—wireless mics, conference phones, control-room accoutrement—that enables communication, but not today, something went wrong. No one can hear anything. Vicky Hipkin is generally a smiley, happy-to-answer-whatever-basis-question-you-throw-at-her type. She's a Scotswoman who emigrated to Canada to work for the CSA. Her fair complexion betrays her nervousness as a blush creeps in and takes over her cheeks. She says, "Today is a big day" in her most convincing voice (which is not very convincing, even though her declamation is warranted).

The mission moves forward in two exciting ways today. The TEGA high-temperature experiment is on the agenda. TEGA will heat the sample to 1000 degrees Celsius and cause some mysterious reaction that identifies organic material. (These are carbon chains—not pesticide-free veggies—that are the building blocks for life on Earth.) That's exciting. This new TEGA experiment could tell us if Mars is habitable.

We're getting ahead of ourselves. Vicky wants to talk about digging.

"There is a revised digging strategy, we're moving to the Wonderland trench," she says. "I'd like Ray to please comment."

"We're moving away from Dodo and Goldilocks," Arvidson begins, doing his best park ranger impersonation. "We'll be swinging over to Humpty Dumpty and opening up the national park system. This next phase is about looking at soil strata." This park system was identified—by best guess—as having the most potential for discovery. The strata, or dirt layers, are the key to untangling the story of what's happened to the polar caps on Mars.

"We're looking to the trench near the top of the polygon. That sample could come from the left of the sample trench in the area called Cheshire Cat," Ray says.

The teams give their morning reports, updating us on everything that happened the sol before that we might have missed. Everyone gets started. We're working toward the new dig and waiting for the results of TEGA.

I FEEL EXTRA CHIPPER AND THERE'S AN HOUR BEFORE MIDPOINT. WITH one complicated topic tackled this morning, I think I'll go for another. I feel surprisingly ambitious today.

I start to learn more about what the instruments really do when they "discover" something. I spent my mid-morning break trying to learn what mass spectrometry is. Don't tell the TEGA team, but I have an ulterior motive: I want to know exactly what it is they do. Then maybe they'll invite me to the TEGA parties. I went to one last year at a training mission. I ended up squished in the back of some half-crazed Hollywood producer's Mustang with a pile of scientists on my lap. I don't really want to get into it any further.

The leader of team TEGA, Bill Boynton, is one of the folks on the mission who makes you say, "I know that guy is awesome. If only I could understand what he's talking about, I bet he'd be *particularly* awesome." He's half the reason we're here. When he gives these full-on TEGA science presentations where everyone seems to be highly engaged, and I twiddle my thumbs, it makes me feel inadequate. What if he finds organic material and I have to answer questions about TEGA on *The Today Show*?

Peter Smith and Bill Boynton will be all "Mars Mars Mars Mars," and the easily befuddled Matt Lauer will stare at them, mouth agape, and then I'll jump in, "Peter, Bill, please let me handle this one." I'm a national hero. Everyone cheers.

This is my moment to tackle the sublime, molecule-bending effects of mass spectrometry. Then I will understand what Bill Boynton is talking about, get invited to parties and be a big star. My ambition is met by good fortune: Mads Ellehøj, one of the Danes who works on TEGA, isn't too busy. He takes pity on me and agrees to explain it like I was a child—a *very young* child.

First, Mads talks about "valence shells" and "atomic weights." I smile politely. I don't want to out myself as an imbecile just yet. That's when I hatch a dastardly fiendish plot. I say, "My audience doesn't really know a lot about these things." I tell him it's for the book. And that he should keep it simple for, you know, "the people." Then I felt the cold wave of morality wash over my drug-ravaged, stop-at-nothing to learn the ins and outs of mass spectrometry brain.

(Yes, I used *you*, dear reader. I'm so ashamed. I'm sorry. I just wanted to cover for my sad inability to recall a single bond from Honors chemistry. Now there's the crushing guilt. It will probably take years of couples therapy to repair our tattered relationship and regain your trust. I hope we can work past this. It's the drugs, I tell you!)

Mads, unaware of my betrayal, is happy to re-explain. All stuff breaks down into elements. These elements each offer a particular essence, their atomic weight. In our strained analogy, their scent. The TEGA is a combination baking and sniffing device—the TAs (ovens) and EGA (gas analyzer). It bakes Mars dirt and then sniffs out the particles in the elements it drives off.

"It's kind of like how you know when bacon is frying up in the kitchen, even though you're in the bedroom, it's got a specific scent, you might even be able to detect that it's eggs and bacon." Mads is keen on using bacon-based analogies to explain science. These TEGA samples are heated in a hand-made ceramic oven material. The gases that evolve (bake off) from the sample are chock full of protons and neutrons. These protons and neutrons are the essential crumbs that make up the elements in the sample. The gasified particles mosey past a complicated array of magnets that are all wired to sensors. Then comes the bit of magic: the trajectory of this brew is measured. *How?* Who knows? Well, Mads does. It has to do with how the particles move past the magnets. The path of their movement is unique. The EGA can read them as little peaks and valleys in its olfactory brain. These peaks and valleys can be matched to a particular element or combination of elements. Once the elements' signatures are identified, you can look for a match, then you've identified what elements are in your dirt. And now you too can become a spectroscopist. Thanks, Mads.

BOY, DO THESE PILLS WORK! CAN YOU SEE HOW MUCH FASTER I'M typing!? It's true. Much faster. Hmm, maybe I can get Cephalon®, the company that makes Provigil™, to sponsor the book. Not only do you type faster but you have great ideas. Lots of fantastic ideas like finding a big-pharma sponsor for your Mars book. *This book is powered by Provigil! For stellar results, try Provigil. Keeps you alert and the taste is out of this world!* My tiredness is gone and I've got LIDAR laser beam focus.

I'm exploding with more great ideas. For instance: NASA should buy a cruise ship that sails around the ocean to keep up with Mars Time—a floating Mission Control. If Virgin Galactic can sell tickets to space for $200,000, NASA can certainly sell out a themed cruise. Discovery during the day; Peter Smith gives lectures in the evening. No more Mars lag, there's a buffet and you can play shuffleboard.

I'm doing great. Except for the odd balloon feeling that's been putting a bit of pressure on my head. It's like the balloon is around my brain but under my skull, and it's about to burst. It's making for a kind of strange disconnectedness. I guess that's the Mars lag duking it out with the Provigil™ in an epic battle of the brain. But no matter, all the weird feelings and strange thoughts can be overcome. I just have to take one more Provigil™ and maybe a Xanax to combat the balloon burst. If that doesn't work, I'll throw in an after-work cocktail and a barbecued steak. It's a fine balance. If I get it right and can stomach a few peyote buttons and a half-naked Indian spirit guide, I'll take it. I've got to go; there's a press conference.

ALONG ONE OF THE BACK HALLWAYS IN MISSION CONTROL, PAST THE MECA lab, is a long narrow room. It's painted black and has a large peephole. It's not where they hide aliens or count the Mars money. Nothing particularly clandestine happens there, yet the sign on the door reads "Get Outta My Swamp" under a picture of the Disney ogre Shrek. I stand there and stare for a while. I can't figure out what Shrek has to do with Mars. I want to understand. Unfortunately, there's nothing to understand. Hyper awareness and expanded consciousness backfire. I've started assigning meaning to random events that don't, in fact, have any meaning. *Yeah, but why an ogre? 'Cause he's green?*

The editorial board meetings and teleconferences happen in the Shrek room. This room is where the media affairs team crafts its Phoenix narrative. There is a lot of non-science talk in this room. Which, of course, is great for me. I'd even participate in the talks if it weren't for Sara Hammond always giving me the "Shhh" finger. I've come to the media room for today's editorial board meeting and telecon. Hammond politely gives me the international "keep your mouth shut" hand signal when I take my seat. I mentally tell her to fuck off. It manifests itself as a polite smile. Maybe it was a smirk. Potentially this passive-aggressiveness is a side-effect of the Provigil™. My head hurts.

There's an open phone line with the JPL media staff, and some NASA folks are on the line, too. Sara does not want them to know I'm here and makes sure I understand with a few more hand signals. You start the morning America's greatest gonzo space journalist and before you know it, your space family is hiding you in the basement.

I like to believe that all these media folks wanted to be astronauts—that they *might* have the unmitigated drive, strength, and intelligence to actually be an astronaut or rocket scientist—and since we can't all be astronauts, they instead chose to toil away in the space business. They know that inherently Mars and the Phoenix mission are a rainbowgasm of awesomeness.

We're sitting at the table with the PR team and a few scientists. Bill Boynton is re-explaining some of his initial TEGA results when I arrive.

"We've done two heating runs," Bill Boynton says. "The first was just about room temperature to see if there was any ice. But we didn't expect that, because the soil has been sitting out in the sun. The second run was about 174c, and at 85c we got some CO_2. Nothing particularly remarkable, yet. Viking saw the same thing. TEGA is working well and that's what I'm most happy about." The media clamors for the results of a high-temperature run.

TEGA's high temperatures might measure organic material. Oooooh, life on Mars! That's what everyone wants, something about organics and life on Mars. That makes a good story. It's a complicated story. As for TEGA working well? Not so much excitement. From your average news-gatherer's perspective, this is an expensive machine we put up there on Mars—it's *supposed* to work.

"I'm not ready to talk about the high-temperature ramp. It will be several weeks, up to a month before we start to understand the data," Bill says. Instant gratification-seeking monkeys that we are, that's not quite what anyone wants to hear. *Cue hemming and hawing.*

We take a bit of a digression and talk turns to problems with the "mainstream media." The phrase "mainstream media" is tossed around with a sort of fist-shaking ire.

The JPL folks think they are more interested in what's going wrong than what's going right. They're particularly annoyed with a *Los Angeles Times* writer who always seems to find the negative in anything they say. He certainly lost his love for space.

"They're always looking for man-bites-dog stories," Sara Hammond says. She's probably right, but that might just be because we failed to explain how it's all actually going right.

PRESS CONFERENCE. RAY ARVIDSON AND BILL BOYNTON DO A microphone level check. Sounds good. We all sit around the long table in the image processing room.

"Bill and Ray, thanks to you guys for doing this," Jane Pratt, the JPL press boss, says.

"We've got a lot going on right now but, hey, this is what we do," Bill says with a sigh. There's a hint of passive-aggression directed against the outreach team for forcing him to ignore his research and waste his time with this whole endeavor. It's possible this press conference affords him a happy opportunity to break from the rigors of finding life on Mars. Probably not. I wonder if he's taking Provigil™, too. Not that I'm insinuating that. I just wonder, is all. This is conjecture based on the inflection of his voice, melding with various complaints about media outreach I've heard, and my inability to process the thunderstorm of thoughts raging through my mind. I conclude he's probably hopped up on Provigil™ and that his hyper-attentive state would be far better suited to digging for Martian dirt than facilitating public relations. Maybe it's just me. I don't know. Paranoia has me in its clutches. He's got to be taking Provigil™. Maybe not. I'll drop it.

But there's a problem with the telephone queue for reporters who dial in. There are far more than expected. This is great! People care!

Thirty-nine participants on the line. That's everyone. We are ready to start. The operator tells us she is going to play some music and then we'll be ready. I get a little nervous. I'm not sure why. I don't have to do anything. Maybe it's time for another pill.

Jane Pratt kicks off the meeting. She introduces Ray Arvidson and he starts to talk about the images of Dodo-Goldilocks trench. He shows the topographic and elevation map; images made using the SSI. The exciting development in the trench is that a there's a nugget of light-toned material. Maybe salt. Maybe ice. Something interesting, that's for sure.

The images are color-coded based on elevation.

"I'm going to talk about the 'red part'," Ray says. This is the top to one of the important polygon shapes. The image shows we excavated from the top of a polygon into a trough.

"We think we clipped an edge of the light-toned material. When we're in the trough, it's deeper or lower or non-existent," Ray says. The "light-toned" material is almost 99.999% certainly ice. Still, Ray hedges. He doesn't want to tell the media they've found ice without proper scientific confirmation. Boring. But he hints at finding something.

"We'll keep looking at it to track the change. If it's just a cold finger we should see frost. If it's an isolated piece we'll expect some sublimation," Ray continues and then lays out the digging plan for the next few sols. "We'll be excavating a new trench." The trenching strategy is important to the scientists but clearly boring for the lay media, which seems to only understand "life" or "no life," "water" or "no water." They want water, ice, or aliens. There are no other stories.

Bill Boynton does his spiel and says how pleased he is with TEGA's performance. What he skips is all the good stuff I'm only *just* learning about: the mess of design process, finger-pointing, a complete rebuild two weeks before the mission, and that, in truth, he's just happy it actually works. The imp in my brain really wants to tell this story. The media, nay, the *world* needs to know how hard this all really is. It only appears simple because without living in Mission Control for a few days it's impossible to factor in all the things that can and do go wrong on Mars, 200 million miles away, that would never be a problem on Earth. Ranting over the phone line will wake them from their Mars stupor. But Sara's chronic admonishing look reminds me to act otherwise.

"We're looking for H_2O, CO_2, and SO_2. But we don't expect to see it at low temperatures," Bill says of the first two heating runs that TEGA has made so far. He finishes up his turn, and it's time for questions.

The *San Francisco Chronicle* reporter asks if they've dislodged white material. Ray shows him the chunk in the image.

A reporter from *The Tucson Citizen* asks if there is water.

"Probably," Bill says. "Water is chemically bound in minerals, which it releases as gas, giving the fingerprint."

The *Los Angeles Times* writer is up next. Oh, no.

"So your predictions are wrong, then, if at 20 centimeters you haven't found ice?"

"Good question," Ray replies, diplomatically. "In fact, we've only dug 5-7 cm, but the bottom of the trench is 20 cm from the top of the polygon. The RA can go 50 cm deep, but it's only 5-7 cm here." Headline: "No water." Geez.

CHAPTER SEVEN

WONDERLAND

SOL 22

IT'S 5:07 A.M. LOCAL TUCSON TIME, 14:22 P.M. ON MARS. SUZANNE Young and Richard Kornfeld sit at the edge of the conference table in downlink, reviewing the satellite schedules for relaying data. They need to precisely coordinate the timing of Phoenix's messages back to Earth. Missing a data pass could cause them to lose a day of science.

"Can we go through it once more?" Kornfeld presses Suzanne. The timing doesn't sit right with him. Suzanne has already done the work and she's taken him through it.

"Don't you trust me?" Suzanne asks, gritting her teeth a little.

"Trust is good, control is better," Kornfeld says. Suzanne rolls her eyes. Kornfeld continues without skipping a beat.

"I love working with him, but somehow when he's on duty the shift lasts three hours longer than what's posted on the schedule," Suzanne says with a wink. "And be prepared for a lot of pressing questions," she adds. There's a method to Kornfeld's madness. He's slow but he gets things done right. For those of you who don't subscribe to *Engineer Aficionado* or didn't recognize his voice from the above chat with Suzanne, Kornfeld is *the* guy who called the play-by-play at landing just a few weeks ago. He's a real engineering celebrity. Now you recognize his Swiss German accent. There's no mistaking that precision cadence and warm staccato that captured the world's attention on landing day. He had the whole world hanging on his every word.

I keep the video file on my computer for the moments I need a little boost—beyond the copious amounts of Provigil™ coursing through my veins. It does far more for the psyche.

Kornfeld has a gift for engineering-inspired gems. If you're giving him some long-winded answer during one of his friendly inquiries, he'll say "I was looking for a binary answer," meaning "yes" or "no."

LANDING DAY OPERATIONS TOOK PLACE AT THE JET PROPULSION LAB in Pasadena. JPL may be wonky with their press, but they're awesome at landing spacecraft. They're really the only game in town. Even the European Space Agency consults with JPL for Mars landings. So they get to enjoy all the glory for that. On landing day, Peter and his landing team were all in Pasadena while we waited back in Tucson. Peter stood in the other Mission Control, the Pasadena-based landing center.

"Fire the thrusters," he'd scream at Kornfeld.

"Thrusters fired, sir," Kornfeld would respond.

"Ah, sir, you do know that it takes fifteen minutes for the signal to travel to Mars," Kornfeld would say after a long pause. It's more fun to imagine Peter yelling at his crew, but landing doesn't happen like that. It's all controlled by Phoenix's computer brain. About a thousand precisely timed events that all rely on one another have to occur in the span of seven minutes. Since it takes about that much time for a signal to even reach Mars, Peter wouldn't get to say much before the lander was already on the ground in a pile. Supervising the collection of data is all there is to do during landing on a mission like this.

"Landing happened ten months ago," Pat Woida said back on landing day. He's so practical. "There's no point in being nervous now." The landing for the most part was a fait accompli after the final instructions were uploaded by the Phoenix engineers. There are several opportunities to adjust the trajectory so Phoenix hits the atmosphere just right, but the last opportunity for a course correction passed hours ago. There's really nothing left to do but sweat.

And in spite of the over-air-conditioned space, I do. Once the landing operations terminate, JPL hands over control to the SOC in Tucson and Peter races back from the JPL campus to re-join the team.

In Tucson, we ate hot dogs and watched a live feed projected onto a massive screen just opposite the PIT. The education and public outreach team hosted a landing-day party for friends and family of the mission. Between pleading with Sara Hammond to let me in the secure area of the SOC, and feeling alone at an amazing party, I'd stuffed my face with wieners and watched the scientists' husbands, wives, parents, kids, aunts, and uncles make nervous small talk.

Then everything got quiet. "Atmospheric entry on my mark," proclaimed the velvety man voice over the PA. *Here we go.*

"Five—4—3—2—1. Mark," Kornfeld counted down. With that, Phoenix dipped into the atmosphere and made its move. I imagined the spacecraft starting its fall: her heat shields deflecting the hot glowing plasma as she plummets through the hazy red atmosphere, beginning a seven-minute super-braking burnout to decelerate from 12,500 mph to zero without any splatter.

"One minute past entry. Still a signal via Odyssey. Standing by for Odyssey data switch to 32K in 45 seconds," Kornfeld updated his team. The bespectacled Kornfeld is officially the Chief Communication Engineer for entry, descent, and landing. He is responsible for how Phoenix communicates during landing. On previous Mars missions, communication had dropped out as the lander went into its seven-minute fall. These moments of silence are part of Mars landing legend. They're called the "seven minutes of terror." They make your mission or your mess. I tried to read an academic paper Kornfeld co-authored that described how it all works. I made it halfway through the synopsis. Kornfeld and his colleagues tried to engineer out the terror from the seven-minute landing sequence. They wanted to at least collect data so if the worst happened they could learn from it. From his wan pallor, you might deduce that terror reduction is a taxing process.

"Standing by for telemetry acquisition . . . parachute deploy detected," Kornfeld says.

Everyone applauds. That's a good sign. Phoenix must release its parachute and then blow off its heat shield. The timing has to be perfect for these events. The heat shield has to protect Phoenix from the friction of descent, but then fall far enough away from Phoenix so they don't make contact at any point along the way.

"Radar switch to altitude mode. Standing by for altitude conversion," Kornfeld announces.

This is a critical moment. We're still waiting for the radar to lock. If it works, there should be an altitude reading to tell us how far Phoenix is from the surface. If it's not working, well, you don't want to know. The radar was a big headache. This is a Polar Lander inherited radar. Yes, the same one that likely caused the Polar Lander to crash. They had to find the problems and then engineer them out. It was a lot harder than they imagined and the radar was still flaky six weeks before launch, making this a particularly tense moment.

"Heat shield trigger detected . . . ground velocity 90 meters per second . . . ground velocity 80 meters per second," Kornfeld keeps making calls. No dropout yet.

"Radar switch to altitude mode detected. Lander leg deployment detected. Radar reliable," Kornfeld says.

A small wave of excitement rushes through the crowd, touching everyone who knows what those steps mean. If they're excited, I'm excited.

"Radar acquisition 2000 meters."

I am a bundle of nerves, frozen with fear, hoping and willing with all my mental power for it to go right. I don't care that this was all programed months ago. I'm nervous as hell. My fingernails are digging into my palms and even my bum is clenched. Peter's ten-foot-tall head and wide-eyed, unblinking face is projected on the wall. Peter points his index finger in the air. At first I don't know what he's pointing at. But then I see it was the aftermath of a premature count-down. He started with five fingers at 5000 meters but after 2000 meters Kornfeld switches things up on us. We expect him to say 1000 meters. But he wants to draw out the drama and goes 1900 meters. Peter is left with his index finger extended. He's just frozen. Desperately pointing at nothing, willing with all his will that the spacecraft lands softly.

"1700 meters . . . 1600 meters . . . 1000 meters . . ."

This is it.

"Gravity turn detected."

Engineers applaud.

"500 meters . . . 400 meters . . . 100 meters . . . 27 meters . . . 20 meters . . . 15 meters."

"Stand by for touchdown . . ."

You can almost feel the pressure drop from the collective inhale.

"Touchdown signal detected! Landing sequence initiated," Kornfeld announces explosively. Peter picks up the small man next to him in a giant bear hug. They look so happy. We cheer, high-five, and some people even cry. Me? There was just something caught in both my eyes.

From a technical perspective, the landing proved flawless. They maintained radio contact during the entire stretch of the "seven minutes of terror." Much of that was due to Kornfeld's communication package. But who cares about all that: WE MADE IT TO MARS!

BACK IN THE SOC, 22 SOLS INTO THE MISSION, IT'S TIME FOR KICKOFF. Vicky Hipkin is sci-lead today. When the team quiets down, she begins.

"This is a historical time in our plan. We should get our first look at a high-temperature ramp. We could even see some organics," she says. Organic material is a kind of Mars Holy Grail. (Maybe aliens are the actual Holy Grail, but this is certainly Dead Sea Scroll territory. Far better than a Holy Tortilla.) Organics are a strong indication that there's life. Finding ice and organics would earn the mission an "A" for sure.

This is Vicky Hipkin's second day as sci-lead. When Vicky arrived this morning, Kornfeld had some words of encouragement for her. He wanted to let her know how well she's done.

"You did a great job for your first sci-lead" Richard said, then correcting himself, "No . . . she did a great job, period."

Even though she's doing great, Ray is antsy and stands next to her, waiting.

"We are on a two-sol trenching excavation. We're going to give the RA team two sols to get us some white stuff," Ray says finally, interrupting. The white stuff is now the focus of the mission. It's become a sort of joke that no one will say what the stuff is. This white-stuff/ light-toned material Ray talked about in the press conference yesterday.

Off the record, it's likely ice. No one wants to be responsible for a leaked news story about finding ice on Mars, if the team isn't damn certain and ready to announce. Retracting a statement like that is embarrassing for NASA and JPL. The outreach team and scientists are highly concerned about embarrassment.

Now that the first TEGA results are in, the team would love to have a little bit of that white stuff to round out their understanding. They want whatever is in the scoop to go down TEGA's gullet. There are eight TEGA ovens but only four wet chemistry cells in MECA. Each sample gives different information about the place they've landed. Ray wants to be sure that they get the best possible representation of the environment to sample.

"We'll look at the light-toned material to determine where we get our WCL sample. And we'll also continue to monitor the Goldilocks trench," Ray says, hijacking the meeting.

"Also, an important note. On [sol] 19 we exposed a chunky bit [of white material] but we didn't hear about it until 20," Ray says. Then he pauses. "Now that chunky bit is buried. In retrospect, it would have been a great discovery to look at. I told the press yesterday we were going to watch it. Now we can't watch it." Ray is a little annoyed, but chalks it all up to a learning experience. Now we need to get some of that white stuff into our instruments and see what this planet is all about.

"Please mention any discoveries early," Vicky says in the same teachable moment tone.

"You get us that white stuff and then we can scrape," Ray says to Ashitey. He hopes that new white stuff will help the media forget about the chunk of whitish stuff he described at yesterday's press conference.

Vicky calls for today's weather report.

"Clear Skies. Nothing new," Palle says. He's a man of few words.

"It's another beautiful day on Mars. See you at midpoint," Vicky says, dismissing the team.

THE FIRST IMAGES OF THE TRENCH COME BACK. A FEW FOLKS STAND around the front monitor looking closely at the images. We all expected more of the white stuff. But there's no white stuff. There's something stranger: black stuff. A new dark, almost shiny, material is unearthed—err—unmarsed.

You want light stuff. Mars gives you dark stuff. No one is making any guesses. Just a lot of looking around and head-scratching. I can't quite tell if it's obvious to them what it is, or if no one has any clue.

"It's hard to be the dig czar," Ray Arvidson says and sighs. Dick Morris comes over and pats him on the back.

"Yes, I'm sure it's tough being a dumpmeister," he jokes.

Bill Boynton, Bob Bonitz, Joel, and Ashitey have a little sidebar about how feasible it will be to get another TEGA delivery by sol 25. They discuss the matter with Dara Sabahi; he's the chief engineer for the Phoenix project and he's here on some kind of secret mission. Three sols isn't a lot of time. The original trench Dodo-Goldilocks had white material. But they don't want to take a sample from there. They want to go into the center of the polygon. That's why they're digging this new trench. Only, this doesn't have white stuff; it has black stuff. Do they want black stuff? Heck, yeah. But they should probably take a moment to observe.

Aaron Zent, one of the MECA co-investigators and a strategic science lead, eavesdrops, too. He comes over to protest.

"They won't be ready with those blocks until sol 28," he says. Blocks are the units of code that make up lander activities. From his strategic work, he knows that there's a bottleneck and they can't just do another TEGA sample acquisition. The debate is tabled; more images come down from Mars. More dark shiny material.

At the big "Jedi" monitor, the water-cooler equivalent here in the downlink room, a crowd gathers. Everyone is a little perplexed.

"Maybe we landed on tar sands. If we have a press release that says 'Black stuff on Mars not necessarily oil,' they just might give us permanent funding," Mike Mellon says.

I ask Ashitey to explain what he thinks is going to happen, now that there's a new kind of material.

"You're still here!?" he says. "You know all the other media is gone. I guess this is gonna be a really big book."

"I'm staying until the bitter end," I tell him.

"Go talk to Joseph," he says. "He's working on the digging strategy." I say okay and start to head out, but then he tells me anyway.

"They want the white stuff," Ashitey says. "This is a test dig to see if we can expose some of the various layers. Once the geologists can assess each layer, we can do some precise digging next to it," he says. The various strata of the soil tell a different story about what the past was like on Mars. It could reveal the type of landscape and if water flowed. "The more precise layers they can take samples from, the more enlightening the results they can get with TEGA and MECA too," Ashitey explains. Then he gives me a look that signals he's got to get back to work.

"Okay, so you think the black stuff is oil?" I ask. He doesn't have time for jokes. He says maybe it's black diamonds and walks away. More likely the black stuff is ice with dirt in it.

AS SOON AS ASHITEY LEAVES, THERE'S AN ANNOUNCEMENT FROM THE TEGA team. Good news: they completed another cycle of heating from the sample and the results are back. Bill Boynton shows a graph of temperature readings from the high-temperature run. TEGA's latest experiment. They heated the Baby Bear sample to about 1000 Celsius, about 1800 Fahrenheit.

"That's the hottest it's been on Mars in a very long time," Vicky says. Still all the numbers don't mean much to anyone outside this room. It's up to the TEGAns to make sense of all this.

"When will we have the happy news?" Vicky asks Bill in her most maternal of tones.

"What we see is very subtle and we will need to do another bake and then subtract from tomorrow's data to tell what we've got," Bill Boynton replies without answering. She's asking him to do a year's worth of science in a few hours. He's not okay with that. Everyone would love to see some organic material in this first sample but Bill won't be pushed into making any overreaching statements without time to collect evidence.

THE MIDPOINT MEETING IS STARTING. A BLURRY PICTURE OF THE NEW Wonderland trench is up on the projector screen. Everyone puts on 3D glasses to see it. These must have great scientific value, but it's hard not to chuckle whenever the science team whips out their cardboard glasses and puts them on in unison. Ashitey is today's RA representative at the midpoint meeting.

"From our perspective, it's a perfect trench. From the dig czar's perspective it's a near-perfect trench. We don't have data yet on how hard the material is," Ashitey says about the newly exposed area. This is what everyone waits for. They dug right down to the hard stuff. Getting there is the easy part. Acquiring some of whatever this stuff is will take more work.

"We still would like better DEMs," Ashitey says, lamenting the missing digital elevation maps (DEMs) for the bottom of the trench. Without these maps, it's going to make getting the hard stuff really hard. Not that they won't do it. It'll just take longer and they're more likely to make errors.

"We will use the torque data to tell how hard, relatively speaking, the surface is. GSTG says it's hard. But that's why the geologists get paid the big bucks . . . to guess," he says, taking a crack at Ray.

"Anything dig czar wants to add?" Vicky asks, preempting any interruptions.

"Failure is not an option," Ray says, in a mock heroic tone. He points to Ashitey. "Give me white stuff tomorrow!"

"Joseph Carsten is in charge of the block. He's a fine engineer and I have the utmost confidence in his ability," Ashitey says.

The meeting is dismissed.

TOM PIKE IS A SCIENTIST FROM IMPERIAL COLLEGE AND co-investigator for MECA. His group helped build the atomic force microscope on Phoenix. That's not what he's here to talk about today. Today, he signed up to give a lecture on the mounting evidence that Phoenix had *already* found ice. Pussy-footing around this light-toned or whitish material talk is silly. Today he's going to present the evidence.

The end of sol (EOS) science meeting is an open forum to present the new ideas on the state of Mars. Scientists sign up for ten-minute talks on a new finding or something new to consider. These are heady, science-rich, technical talks that are impossible for non-specialists to follow. But that's okay. EOS is an opportunity to simultaneously flatter science team members while they explain the details of their work so a small child could understand.

"Your talk was fascinating. Could you clarify what you meant in the part after you stood up and said the title of your presentation until the part where you said, 'that's all for today'? I had trouble understanding that section." Without a forum like EOS, Peter and some of the other mission planners worried there would not be enough collaborative effort on Phoenix.

"The risk is that they squirrel away their data," Peter likes to say. It's difficult to keep track of everything that's going on in the SOC. There are so many aspects to the mission and its data, keeping tabs on them all is a full-time job.

"It's also difficult to force people to collaborate," Peter says. EOS is a means to prevent the scientists who lead various groups, the co-investigators, from hiding the messy unclear results until they write up their formal papers on the matter. EOS should be the place where chemists find out about what the geologists think and the atmospheric guys can coordinate observations with the physicists. It's all in the name of interdisciplinary science. All Peter asks is that everyone share what they're up to with the group. Peter thinks it's beneficial for everyone to understand the science thread pulling the whole mission. There will like be fewer conflicts later when resources get scarce and teams start to fight it out for their experiments.

This idea of transparency isn't universal and I've already seen small protests from those not in the same philosophical camp.

"Getting credit is important if you're in that business," Nilton Renno says. How you get credit divides the Phoenix scientists into two ideological camps. The first group includes those who think sharing data and guessing is useful. The idea that information-sharing makes for better science is in line with the democratic spirit of a tax-funded mission. The second camp begs to differ.

"Scientists get scooped all the time," Mike Hecht says. Touchy-feely sharing sessions are anathema to these mission-hardened planetary academics—a recipe for embarrassing stories leaked to the press and your innovative ideas stolen. This second group thinks that since they've done the work, built the instruments and equipment, they should be rewarded with the time to look at the data and the results from their experiments without the fear that it's going to be pilfered by science-scoundrels waiting in the wings. Although I'm inclined to agree with Peter and the first camp, mostly because I wouldn't be here without his lofty notions of transparency, some of the scientists in the second group make a compelling argument.

"There is almost no time on a mission like this, especially where everyone is doing double and triple duty, to do a lot of analysis. The work will take years," Jim Whiteway, who leads the LIDAR team, tells me. "We spend a lot of time building these instruments. It's only fair we get some time to do research in the lab. It's generally bad science to put out random thoughts. These thoughts are better collected into a coherent story that you commit to and publish."

"Scientists aren't used to being rushed," Bill says. "It's more a slow methodical approach." But on Phoenix they only have a short window before the data goes public. That's part of the contract NASA has with Peter Smith. Six months after the mission is declared over, the data sets become public domain. There's limited time to get results before the competition comes wading in. Sharing results and data opens the pool to new thought, and new thought is good.

The injustice that films like *Armageddon* or TV shows like *CSI* do is to convince us non-scientists that scientific discovery and problem solving happens on a tidy timeline. In the real world and on Mars, problems are not discovered, understood, and solved between the opening and ending credits. The "Eureka!" moments generally come after years of analytical work and following lots of false leads along the way. Since our time is short, EOS should lubricate the gears of discovery.

When the sun sets on the lander, there will be no do-overs. Untested ideas or half-baked theories are encouraged. To avoid the inherent conflict baked into the process, these sessions are closed to everyone but the team and an occasional, announced guest. No one wants the media nosing around for crazy ideas about Mars at the EOS. I always sit at the back during EOS and never say a thing.

"This should show that we already have enough evidence," Tom Pike says, making his case that Phoenix has already discovered something wonderful and there's no need to hide behind the coded language of "light-toned" material. "This is the trench on sols 20 and 21. We see two bits of the white material," Tom says. The bits have little measurements on them. Tom shows how the two chunks shrunk over time. There aren't many images in his flip book, so it's a little shaky. To bolster his argument, Tom calculated the rate of shrinking.

"They shrink by 2.5 mm a day, about 100 micrometers an hour," he says. There is a lot of tricky math. When you get lost in these talks, it's important to follow the inflection so you know when he's getting close to making a conclusion.

"This is the rate we would expect sublimation to occur: 1.6 cm to 1.1 cm," Tom declares. He stands proud. The presentation makes it clear that Phoenix has observed ice on Mars. Case closed.

"Ah, which ones are these?" Mark Lemmon asks, unconvinced. "It looks like there is an aspect ratio change. I think that one might be a left and the other a right image."

Lemmon suggests Tom did not prove there's ice on Mars. Instead, he measured the "camera 1, camera 2 effect." That's not really a science term; it's the *Wayne's World* version of the parallax view. It's when you look at the same thing from your left and right eye and it looks different. Try it. It's fun. You can make the words jump around on the page. Camera 1 . . . camera 2. Anyway, the SSI is built the same way. That's how it makes 3D images. But you can't measure something on a picture if you use the left and then the right eye of the SSI camera on Phoenix. Mellon thinks that the melting that Pike sees is actually just different angles.

"We need to be very careful about looking at the same time of day and image if we are to monitor over time," Doug Ming says, shaking his head. He whispers something to a colleague seated next to him. Doug gets prickly in these meetings when he thinks someone is making a claim they can't back.

"Maybe there are some multi-spectral spots of these images?" Carol Stoker, who is a bit more understanding, asks. These images could provide evidence for or against the two-camera effect. Not that any of the comments matter. Luckily for Tom Pike, you don't have to prove anything at the EOS meeting, just state your case for people to think about.

"Thank you for listening," he says and sits down. The grumbles persist for a bit.

Sara Hammond stands up in the back of the room.

"Can I make a short statement?" she asks. "Now I'm just the messenger, but there is a perception that the news is bad from Tucson . . . but we all know it's not." Sara says JPL is upset about the bad press Phoenix is getting. This probably has to do with the "First try finds no water" headlines that came out of yesterday's press conference.

"Please be more careful with statements that might be construed as negative," she says. All the scientists underwent "media training" and Sara has helpfully tacked some of the key bullet points to her door. "There is no off-the-record" is right near the top.

"We're not just taking tourist photos!" Mike Hecht says in an outburst. He has a point. Even if he words it in a slightly misguided way. Tourist photos are one of the most important ways to connect people with the mission they're funding. But that's not his point. What he *means* is that good science takes time.

"We need more time to make sense of what we have. The press conferences should be scheduled when we know more," Heather Enos says in defense of the scientists.

The situation is kind of confounding. It's not the science team's job to craft the story. They're just supposed to do science and fight about results. Telling the group they're mucking up their own story is certainly not much of a morale booster. Sara leaves. And the short-term planning session begins. There's a long night of strategic planning if they're going to get the hard stuff in TEGA any time soon.

CHAPTER EIGHT
THE LOST DAY

SOL 23

CLIVE COOK SITS IN ASTG LOOKING ANNOYED. HIS COLLEAGUES twiddle their thumbs too. Seems as though something has gone wrong with Phoenix.

"It didn't work. You can quote me on that," Clive says.

Somehow yesterday's plan failed. There was no TEGA high temperature run, no digging, and no atmospheric data collected. The details of the failure remain a mystery. Not just a mystery, but an international mystery. And since Clive is from the U.K. and works with the Canadians, he can't say what exactly the mystery is. And no one is allowed to tell Clive, since his status as a foreigner puts him under the jurisdiction of the International Traffic in Arms Regulation (ITAR), a federal law that restricts the export of military hardware. So even though he looks normal, Clive is an ITARd. Some people feel more comfortable saying "ITAR restricted." Because of his ITAR status and the fact that no one will tell him what's going on, we all infer that something is wrong with Phoenix's "brain."

Clive works in Jim Whiteway's lab in Toronto. Whiteway's team at York University, along with the Canadian Space Agency, a company called OPTECH, and an engineering firm called MDA, built the LIDAR on the Phoenix weather station, MET. Since the MET team is mostly foreign nationals, they're restricted in what they're allowed to know about the spacecraft to protect national security. What does national security and military hardware have to do with our lovable little lander named Phoenix, you might ask? I could tell you, but I'll have to see your passport first.

To get to Mars or anywhere in space, you need to hitch a ride on a rocket and you need software to control things in space. The logic for creating ITAR goes something like this: since rockets carry thermo-nuclear weapons, robot arms, and general science junk equally well, the regulation lumps them together. Even if our intentions on Mars are peaceful, the "Man" doesn't see it that way. The ITAR rules say that uncleared foreign nationals—"FN" is how they're denoted on their security badges—are NOT allowed in flight hardware discussions. Specifically, the 1999 law transferred the regulation of harmless space activities from the Commerce Department to the State Department. It says no U.S. citizen can transmit any harmful information to non-U.S. citizens. "Heck to the FN scientists" is probably how they put it in the bill. It says all sorts of other stuff, too. The practical part for Phoenix is that every time you want to tell an FN a bit of information that might fall under ITAR, you need permission. Getting permission isn't always easy. Some of the scientists have information-sharing agreements for some aspects of the mission, but these are narrow in scope. Since the ITAR regulators at JPL could face severe jail time if they violate the rule, they're a cautious group. "Request not approved" is most likely the response you get when you ask if it's okay to let the Canadians know what's happening. Jail is a huge disincentive, eclipsing the spirit of cooperation by a long shot.

It's not just the social stigma of being an ITARd. This well-inten-tioned rule leads to real problems. The ITAR rules even prevent the MET instrument—not just the scientists—from gaining access to parts of the spacecraft. The MET machinery isn't allowed to turn itself on or off. It needs Phoenix to initiate the commands. The MET and its programmers don't have access to commands like "MET on" or "MET off," since that requires it to talk directly to the spacecraft. While building the MET, team members invested lots of time investigating strange results from a pressure sensor. Something was very wrong, but it didn't make any logical sense.

"They wouldn't give us a layout of the lander deck," Carlos Lange, a MET co-investigatory says. Although it's not clear why that would need to be classified.

"It's a complicated rule," Peter says. "They couldn't tell them the interior contents of the electronics boxes without being in violation, and that caused some issues." The result was the MET team didn't know that there was a heater on the electronics board next to their sensor. Pressure and temperature are intimately connected, so any time the heater kicked on, their results went berserk. Luckily, they discovered the problem before launch, but the instrument was left with limited functionality.

"I couldn't take it anymore," Carlos says. Instead of just drinking from the fire hose of frustration, Carlos devised a scheme to legally circumvent the regulation. Carlos is an outspoken and exuberant professor from the University of Alberta. He is tall and lanky with a heavy Brazilian accent. Every morning Carlos prepares his traditional mate tea from a giant sack of leaves and drinks from a carved wooden chalice and metal straw. Carlos discovered Phoenix after hitchhiking a ride from Nilton Renno at a science conference.

"The rental cars were expensive and I heard Nilton speaking Portuguese, so I waited for him and asked for a ride," Carlos says. His thrifty ways paid off. Nilton needed an enthusiastic and sharp scientist like Carlos to flesh out the team.

Once Carlos was on board, he came up with some initial concepts for the telltale and some math models to understand how temperature and pressure fluctuated near the spacecraft.

"I had to wait two years to get my clearance," Carlos says. This was so he could see the layout of the lander deck to model some of the environmental interactions. They don't have many chances to work on Mars projects at the University of Alberta in Edmonton, so Carlos wanted to involve his students. Plus, he needed the brain power to help him characterize—the sciency way of saying investigate and understand—the effects of the Phoenix heaters on the local environment. The ITAR monster said, "not so fast, you FN scientist."

"My students were not cleared and we couldn't get the fabricators in the machine shop clearance to build any models," Carlos says. So he devised an extra credit project to one of his classes to try to solve.

"I called it the Open Lander Project," Carlos tells me. "The rules were that my students . . . they would have to use the Internet to try to figure out the dimensions and location of all the instruments on the deck and where they were likely located." To make sure they didn't run afoul of the rules, Carlos enforced several restrictions. "Under no circumstance could they ask anyone with restricted information for clarification." So they didn't. Using public domain photos posted by NASA and JPL and hosted by Google, Carlos's students employed the sudoku method to create an accurate model of the lander. Carlos proved that State Department's ITAR rules—as written—are silly and counterproductive. Red-faced, they repealed them immediately. Just kidding, it changed nothing. That's why Clive Cook can't tell me what's happening at work today. His screen is blank save for a little note that says ITAR restricted.

RICHARD KORNFELD CALLS THE TEAM TOGETHER FOR A BRIEFING. He describes some problem with the lander's memory system that's making it impossible to do much of anything right now.

"Too much engineering data was produced . . . there was almost a time out," Kornfeld announces the news like a breaking bulletin on a tragic local event. "The problem is not fully understood. So, we can't save anything to flash [memory] at the moment. Are there any questions?"

"Did we lose the whole day? And do we need to redo sol 22?" Pat Woida asks.

"I defer to Doug Ming," Kornfeld says.

"Well, I'll defer to Jim," Doug says.

With no one left to defer to, Jim Chase is where the buck stops. As TDL, he is the manager of all the bits returning from space.

"I'd like to pull up the chart that shows what data has been lost," Jim says. It takes him a second to find what he's looking for. He turns on the projector. The screen is blank.

"That's what happened to our data!" Peter Smith says.

Chase fumbles for a bit. We all wait while he looks through his folders and pulls up a new chart. Data is the lifeblood of Phoenix. The team uses a set of identifying tags called application process identifications (APIDs). These are the markers that determine how information is prioritized as it comes down from Mars. It's the process for asking Phoenix for something and then finding the answer back on the ground. It's a straightforward, million-step process outlined in the end-to-end data management (EEDM) documentation. I'll take you through it now. Please grab a paper bag if you're prone to hyperventilation. The only thing that gets me through it are the mind-altering Mars drugs.

The science team makes a request to Phoenix to generate a certain kind of data product. Those requests are processed by the relay orbiters and sent down to the spacecraft. The spacecraft executes the commands it receives from the relay orbiters and sends them back as a particular kind of data product. Data products are images or results from an experiment like a TEGA bake. Managing all that back-and-forth is a hot mess.

The computer on Phoenix has two kinds of memory, 100 megabytes of regular flash that's on any old thumb drive and then this other kind of temporary storage that gets erased and re-written. Once the data stored on this second type of volatile memory is gone, it's gone forever. The application process identification (APID) is the special instruction that comes along with the request for data. The APID tags tell Phoenix where to store data—in flash or volatile memory—and when to send it back to Earth. It also has a little note for when it is safe to delete. If you don't get the data you asked for, you can launch an investigation to find out where along the chain it's gone missing. Did the activity occur? Is it stuck on the relay orbiter? Was it deleted? There are plenty of chances to lose your science data. APIDs determine if and when you get your data. APIDs get fun names like "key" or "crit_key" or "crit." It's the tactical downlink lead's (TDL) job to track all that data. If you get it wrong, your one-time-only non-repeatable experiment could be lost forever. Now something screwy is happening with them. Apparently, only the highest priority APIDs were downloaded. Very few bits of engineering data arrived on Earth.

"It looks like the RAC images before the surface dig are lost. We can't get those back. But we'll move on and hopefully it's not a terrible loss," Doug Ming tells the group.

To recap: something has happened to the lander that caused the whole spacecraft to go into safe mode. Safe mode is like the fetal position. Phoenix is a bit of a nervous spacecraft. It uses the fainting goat approach, folding over and taking an extremely defensive position at the first sign of danger. Now Phoenix will wait for the science team to tell it everything is okay. When everything is back to normal, it gets up on its own and goes on working. This sol's plan was not executed and a day of science was lost. Some data from the previous day is missing or lost. Whatever went wrong is disrupting the flash memory. That means no data can be saved. All the activities they do now will have to be completed and then transmitted back to Earth and there's no memory back-up if the transmission fails. This will limit the number of activities the team can accomplish each day.

"Anomalies happen, but I want to show you this," Doug says to the team. He opens a document with a grid that shows everything Phoenix must do to be considered a success. There are quite a few boxes checked. Many are the obvious ones, but still important. Land safely. Check. Acquire first sample. Check. "We are still marching along down the path to a successful mission."

I also like to put stuff I know I'm going to do on my daily list to make me feel good. Eat breakfast. Check. Put on two socks. Check. These ticked-off boxes remind us that we're doing a lot of great work here.

"I'd like to acknowledge the spacecraft team for identifying the [memory] issues very quickly and stabilizing the craft. It's to their credit that we have the opportunity to do science today," Richard Kornfeld says to the group. There's a round of applause. Then it's back to business.

Bob Denise, a senior engineer who built a big chunk of the software that controls these interactions, volunteers to help explain after I stalk him for a couple hours.

"I'll try to explain it to you without violating ITAR rules," he says. I tell him not to worry, I'm a citizen. He's not interested. Better safe than in a federal penitentiary. "The computer that tracks on-board data generates its own data," he says. "Something is wrong with the software. It keeps recording the same data in a loop that it was generating for downlink." The information downloaded by the lander is organized by APIDs. "This key bit of engineering is the highest priority, lowest quantity data on the spacecraft. Since it's the highest priority, it gets downloaded first. There was so much of it, it took up all of the downlink," Denise explains.

"In the short term, we can adjust the APIDs," he explains. The team can code the data that has been causing problems as "low priority" so APID will not load it until later in the cycle and won't ruin what everyone is currently working on, but that will just be a band-aid.

Denise says it became a serious situation because the volume of data increased the boot time of the spacecraft. That's how long it takes for Phoenix to get up and ready in the morning. This time is closely monitored with regular check-ins. Are you ready? I'm coming? *Are you ready? I'm coming, damnit.* They call these check-ins "heartbeats." Phoenix was so busy processing all the data that it took an even longer time to boot up.

"Then it came dangerously close to missing a heartbeat and changing sides," he says. If it misses a heartbeat, the computer assumes something has gone very wrong and moves to backup mode. It switches hemispheres in its computer brain.

"The other side of the spacecraft is like a spare tire. Once you use it . . . well, you don't want to use it," Denise tells me. Because of all this, the flash memory where extra data is stored when the spacecraft goes to sleep does not work. The mission will hobble along for the next few sols, possibly longer. This is annoying but not catastrophic. It is also about the limit any human can take for talking about APIDs.

DARA SABAHI IS BACK IN THE SOC TODAY. SABAHI IS THE YODA TO PETER Smith's Luke Skywalker—a mythic figure that only appears in the SOC for brief moments of consult at critical junctures. He is the chief engineer for Phoenix. If he's here, Sabahi paces around the perimeter of SOC and swoops in to aid any troubled group. He rubs his head and kneads his chin before approaching a subordinate and whispering some prescient advice in his ear. He has a broad build and a formidable widow's peak. His arms are always folded and slumped; he's easily positioned to scratch his head or engage in other classical deep-thought postures. He wears a photographer's vest and camping pants that zip at the knee. We can only imagine what he keeps in all those pockets. In spite of his intimidating and sometimes dour look, it's important not to judge an engineer by his vest.

"You should be pushing for better access," he says after asking me probing questions. *Wait, I'm supposed to be pressing you for information! This is awkward.* He is so good at quickly getting to the heart of a problem. "Documenting the mission will be very important for the future. I am always looking for a record of the last mission when I start on a new one. This is how we learn," he says. I suddenly feel like I need to be a better person. *You complete me, Dara.* I want to tell him that I don't think I'm doing what he thinks I'm doing.

"I'm counting on this documentation," he says. This is just getting worse. "Don't worry. Peter will make it happen," Dara says. And while I feel very small, it's the first time I feel part of the team. It doesn't matter that my paltry brand of investigative journalism is out of scope for him. He's deduced and attacked the issue.

"The more people can read about the mission process, the more we can learn about improving the process. You need to be close to the action," he tells me. "And the action increases as you get closer to the 'sausage grinder.'" The sausage grinder is how he describes the process of turning the scientists' wishes into an actual plan the lander uses. These are the highly technical shift II meetings where tensions run high and young observers get confused. I don't want to get kicked out because someone misplaces their anger and complains. Dara tells me not to worry.

"You know, NASA is actually an open culture," he says. "But emotion can get in the way." He suggests I get into the anomaly response meetings. I say I'm not sure about pressing my luck yet. I want to be here the whole summer and I don't want to step on any toes. He says with confidence that Peter will help. He doesn't know Peter takes more of the tough love approach with the people he finds to document his mission.

Peter needs to tell the anomaly leads that I'm okay to go in, he says.

As chief engineer, Sabahi's job is pretty easy: make sure Phoenix worked. He has the worry lines to prove it was stressful. Before I ever talked to him, I saw him as some JPL apparition to be feared. Sabahi is not just an engineer of machines, but of man too. He talks about the problems of making a mission like this work when so many smart people come together with different agendas. He says there is only one way to do it.

"We create a very stressful situation to train them," he says. These are the training missions all Phoenix ops staff attended. Stressing the team sounds cruel.

"That's how you get smart people who are all used to being right to work together. You have to tax them. Then they can pull together as a group and support one another," Sabahi says.

He understands how large-egoed, big-brain folk are motivated.

"Most people are afraid to confront risk—they don't understand it—because they haven't faced it over and over again. A mission like this is about managing many points of risk. In situations like Phoenix you have risk. But if you don't confront it and move on, you just make it more likely that another issue will come up or the sun will start to set [and the mission will be over]." Dara then starts to levitate because he's about to reach aerospace nirvana. At least that's how it happens in my mind.

Sabahi is one of the few people who sees the entire space production chain.

"From the newest technician to the heads of departments, I see how people work," he says. "It generally goes like this: selfless dedication from the technicians and gutless decision-making from the top." Those who make the gutless decisions don't understand risk.

"That's the problem with the people who make those decisions, they are only afraid of embarrassment. They don't have the courage of character to understand and explain," he says. Uh-oh, I think I'm one of the people afraid of embarrassment. Damn.

"For instance, Pathfinder is seen as a failure from the JPL perspective. The whole world feels like it was a success but because some rules weren't followed it's considered a failure. These are political decisions. That's why I'm burned out and need to get away." Sabahi says that as soon as he's done here, he's quitting. He's going far away. This doesn't just strike me as sad; it kind of makes me worry about the fate of humanity. I might be negligent in tossing around the word brilliant when describing folks in Mission Control, but when the guys I casually describe as brilliant describe Sabahi as the most brilliant guy they know, then you corroborate the fact that yes, he's brilliant.

"They pushed Dara Sabahi too hard," Peter says. "He is great at what he does, but they risk losing him." He got Phoenix to the ground safely, but it came at a high personal cost.

Phoenix is a sinkhole. Addiction, burnouts, breakdown, cancer, divorce—and that's just the "A through D" file. Phoenix will take everything you have to give. She's a remorseless, soul-sucking machine. For example, the simplest part on Phoenix is the bio-barrier. It's a bag that protects the robot arm from contamination, and that little cover-up comes with an 18-page document detailing how it might fail. If the bag breaks, snags, bursts, or busts, the mission would be finished. That's just a bag. In the landing sequence alone, there are more than 200 points of failure that had no backup systems. If any one of these goes, the lander goes. In trying to quantify the risks associated with these points, there's a strange Faustian statistical process. Let's say you have a 2% probability of failure. It might be extremely expensive to get that down to 1%.

"You might tell yourself we should not spend the money to reduce the probability by 1% because that's not very much. But if you examine it from the failure side, moving from 2% to 1% means cutting your failure rate in half. That makes a big difference. So how do you decide when things are safe enough? They never are. And it can overwhelm you," Peter tells me one day, trying to explain why missions are so expensive and complicated. You just do your best to understand the risk, then you hold your breath and cross your fingers.

IT'S 9:19 A.M. LOCAL TUCSON TIME, EARLY EVENING ON MARS. I'M learning to deal with risk and feel rejuvenated after my long talk with Dara. No, I'm more than rejuvenated; I'm inspired. I wander around the SOC. Everyone is busy and stressed out. But I'm confident they'll work through it.

Today there is an open house. It's an opportunity for the public to sit in the briefing area. There will be lectures, a chance to meet a scientist, and it comes with free snow cones at the end. Now that I'm becoming a real insider, I decide to shake things up and visit the open house with the rest of the hoi polloi—err—space fans. I take my seat with nearly a hundred other locals who have turned up for Carla Bitter's early morning presentation, which begins with an animation of the landing sequence. Maybe I'm a little sentimental, but I can't not think about how hard Dara and all the engineers worked to make this happen. The people in this room have no idea that what they're seeing is impossible, and only through the hard work of thousands of people working together and a bit of luck can you make the impossible possible. I want to shake them and then remind them that even getting within a 10 kilometer radius of the landing site requires more precision than hitting a golf ball in Washington, D.C. and get a hole-in-one in Sydney, Australia! And there are very few people who even have the balls to do it.

"Welcome!" Carla says to the space fans. She describes all the elements of the mission, the sample trench, the naming conventions, how a sample gets from the trench into Phoenix. There are pictures and diagrams and animations. The visitors—young and old—applaud and ask questions. Then we eat dirty ice slushies.

CHAPTER NINE

MISSING PIECES

SOL 24

WITH JUST THREE MINUTES UNTIL THE DROP-DEAD UPLINK time (DDULT), the sequencing team pushed their plan out into space. That's the closest they've come. Three minutes later and the Mars Reconnaissance Orbiter (MRO) would have flown over Phoenix and remained silent. Raising its SSI camera eyes to the heavens, the little lander would have put up its robot paw and said sadly, "Why have you forsaken me?" Thankfully Phoenix got its plan and was never the wiser. The final moments of a plan are always exciting, but this would have been one to remember.

If you recall, the organizational scheme of the Phoenix planning cycle is broken into shift I (downlink, tactical planning, and strategic planning) and shift II (turning the tactical plan into robot code). Tactical planning is for the activities you'll do the next day, and the strategic planning is for all the activities you're going to do after that. The day is organized as a sort of "Y"-shaped structure. Shift I builds the strategic and tactical plans, and shift II implements these plans. The two teams form the branches of the "Y."

Shift I workers begin each evening (Mars time) by planning for the next sol. They use any new data to make sure what they're doing is consistent with what's new on Mars. That's kickoff and the data downlink part of the day. The science team then works with engineers to codify these activities into a tactical plan. The two teams come together to discuss the plan and make an official handoff at midpoint; the midpoint meeting marks the break in the "Y"-shaped day. It's where shift I and shift II come together to coordinate the science plan and begin the process of converting scientific activities into lander instructions. After midpoint, the shift II team takes on a tactical role preparing the day's plan, while shift I moves on to planning the next few sols.

Implementing the plan, the shift II tactical role, is a difficult, detail-oriented process that must be completed before the last communication pass of the day. If they can't complete the plan, the science day is lost. The key to a successful midpoint meeting is that all the instrument sequencing engineers (ISEs) and shift II personnel are comfortable with the intention of each activity in the plan. This is what makes the 80% rule important. The engineers must be 80% confident that they can complete the code for the activities within an eight-to-ten-hour window. It's a funny construct, but that's how you get good results with imperfect information.

The engineers are artisans, hand-coding to the lander program in a special robot coding language called virtual machine language (VML). Each activity is a sequence. Occasionally there are sequences that are already tested and used frequently; these are blocks that live within a library of activities. The coding process has lots of rules to keep errant commands from damaging the spacecraft. The guardians of these rules, mission managers like Richard Kornfeld, Bob Denise, and Julia Bell, are the spacecraft's protectors, and they take their work very seriously.

Even after all the code is tested and compiled, the instrument sequence engineers (ISEs) are asked to describe for the group the riskiest aspect of the code and how the code they've written might fail and what they've done to prevent that. Once the mission manager feels satisfied that the code is solid, then the shift II lead and the mission manager II must sign paperwork authorizing the plan. They take the authorized plan into one of the few places in Mission Control—besides the women's bathroom—where few people are allowed, the spacecraft room. From there, they fax the paperwork to a man called the ACE. (After hours spent researching what exactly ACE stands for, I learn it's just a name. The guy who directly commands the spacecraft is the ACE.) The ACE is then authorized to radiate the plan through the terrestrial satellites of the deep space network and on to Mars.

I'm glad they made it in time. Losing another day, just as things get going again, hurts.

TO GET A DIFFERENT PERSPECTIVE ON THE MISSION, I GO FLYING OVER the desert. Nilton Renno set up a flight with his friends at the Tucson Soaring Club. Nilton is an accomplished sailplane pilot. He used to race until a friend died in a sailplane accident.

"I have a family, and those risks no longer seemed appropriate; not for a sailplane record," he says. "But seeing these processes in action is a big inspiration for my work." Nilton actively studies the convection and atmospheric forces that make it possible to glide. Gliding gives you a chance to experience the interactions between the ground and the atmosphere. There's no better way to get the visceral impact of physical processes and the silence to contemplate them. I couldn't resist a chance to interact with the atmospheric forces that inspired Nilton to become a scientist.

NILTON RENNO AND PETER SMITH COLLABORATED FOR MORE THAN A decade.

"It was serendipitous," Nilton says. "One of my undergrads came to talk with me about dust devils, and I didn't know much about them. I told him to come back and we could start to review the literature on them." Dust devils are the swirling masses that look like tornadoes but form under clear skies. Researching only drove their curiosity.

"There was no comprehensive theory to explain them," Nilton says. Curiosity turned into an obsession.

Meanwhile, on Mars, Pathfinder and Peter snapped photos of dust devils. Then one sunny afternoon, Peter and Nilton got to talking and found they had a similar passion for swirls. After the Pathfinder mission concluded, they collaborated on a project called Matador.

It was a typical bromance. A couple dudes finding a common love of Martian dust devils. Their Matador project was a massive investigation of how dust devils affect the atmospheres of Earth and Mars.

"We built the instruments for the cancelled '03 mission," Peter says. "Instead of using the equipment to learn about dust devils on Mars, we brought it out to the desert and learned about dust devils on Earth." When it came to for Phoenix, it was only natural that Peter would ask Nilton, his colleague and the resident atmospherics expert, to lead the MET team when he first conceptualized the Phoenix mission.

"In fact, I was in Peter's office when Chris McKay and Carol Stoker called to discuss a mission with Peter as the P.I. After the call, I was really excited. I said something like, 'Let's do it!' Peter then asked me to lead the MET instrument."

At first, Peter and Nilton decided to ask their colleague from the Matador project Allan Carswell to build the LIDAR for MET. Carswell is a Canadian science legend and LIDAR expert. Partnering with Carswell and the CSA (Canadian Space Agency) meant they would get LIDAR expertise, and the CSA would cover the cost. Those savings could mean the difference between making or breaking the cost-cap on a super-frugal mission like Phoenix. After NASA shortlisted Phoenix and the prospect of being a real mission grew closer, "philosophical" differences emerged.

"When Nilton accepted the project, he was here at the University of Arizona. Then when he moved to Michigan, the University there was eager to participate, but things got more complicated," Peter says. There was a disconnect between Nilton and the proposal manager at JPL. "First there were contractual issues, and then Nilton's costs kept going up." Nilton's lab and JPL couldn't come to an agreement on costs. They claimed that Nilton was difficult. Nilton says being "difficult" means not agreeing with their accounting. Words were exchanged. Trust was broken. In the end, JPL awarded the entire MET instrument development to the Canadian Space Agency. Peter asked Nilton Renno to remain a co-investigator and lead the atmospheric science theme group.

"I felt a little deceived," Nilton says. "But I was happy to focus on the science and lead the group."

"I don't think there was deception, just a matter of controlling cost," Peter responds. "These were hard decisions we had to make.

The scientists are invested in their instruments, and there aren't many opportunities to go to Mars. Sometimes there are hard feelings. There was another instrument we were considering but had to cancel. The guys who were going to build that instrument still turn away when I see them on campus."

The move saved the mission a lot of money and offered a chance for NASA to include Canadian partners. For Nilton, it was a chance to focus on the big questions.

"As ASTG theme group leader, I spent my time working to formulate science questions, developing science requirements, and working with all instrument builders and co-investigators," Nilton says. And the work paid off. "People frequently mentioned that the ASTG was the most well organized and efficient theme group." Before the launch, "we at the ASTG discovered that ice scraped by the RA would sublimate faster than could be collected to be delivered to TEGA; this led to the addition of the RASP drill to the RA. We proposed the telltale, we helped refine the sample delivery, and much more." Freed from the time-consuming contractual obligations of delivering parts and paperwork, Nilton had a bit more time to sit back and think. For Nilton, it turned out to be a gift. He had time to do his best thinking. And for Nilton, his best thinking is done soaring over the Sonoran Desert in a sailplane.

Most of the scientists on the mission work on a particular instrument. It's time-consuming to build equipment and design experiments for Mars. On larger space missions, there are scientists dedicated to instruments and the big picture. For this mission it wasn't possible. There wasn't enough money in the budget to have lots of people doing this kind of big-picture integrated science. Instead, the mission was structured with four theme group leads that would make sure the instruments development tracked with the science goals. And then, during the mission, they help Peter align the Phoenix goals with all the new evidence and experimental results. The streamlined approach works within the budget constraint. Unfortunately, this streamlined structure limits the time many of the scientists have for just doing science.

"That's why we come in on our days off," Suzanne Young says. "When else would we have time?" For the scientists that have double duty manning instruments and working on the daily plan, the science analysis they came for can seem like a luxury. The four theme groups are supposed to ease that process, but they introduce other problems.

"We have to analyze the data as it arrives each sol and test hypotheses developed on the fly, otherwise you might never be able to test them," Nilton says. Nilton is one of the few who has time to do this. He spends his sols devising strategies and testing new hypotheses. Whether he's got an ionizer cranking on his desk or dropping things and timing them, Nilton likes to pursue lots of avenues. His free time doesn't make him a lot of friends. During development, many saw Nilton's tinkering as meddling. They weren't sure who had appointed him mission gadfly. He seems to wear that epithet as a badge of honor.

"It's what happens when you're very honest," he says. "But at least you sleep well at night."

Before launch, Nilton had a hunch that interesting things might happen under the lander.

"I wanted to see how the rockets would disturb the landing site," Nilton says. The lander used pulse-rockets to touch down on Mars.

"My lab thought probably it would excavate a lot of material," he says. That would have uncertain impact on the pristine Martian environment. NASA puts extraordinary effort into planetary protection for Mars. They relentlessly clean the spacecraft so they don't take any bacteria or spiders to Mars. But what happens when you pound the surface with hydrazine-powered rockets? No one knew.

With some of his graduate students, Nilton went to work on a series of thruster-plume experiments.

"Manish, one of my graduate students, set up a slow-motion camera in one of NASA Ames's Mars rooms," he says. Nilton used a special 16,000-frame-per-second camera (your normal camera shoots about 30 frames per second) and a glassed-in, Mars-like sand pit to set up his "Mars-Fire II: The Return of the Nilton" shoot.

The film, in super slow-mo, shows a tsunami of flying topsoil cleared away by violent shockwaves. "It was an interesting result," Nilton says. The thruster plume dramatically heats the surface and even turns some of the sand into glass.

"The French Atomic Safety Commission invited me to give a talk about the results of this test." The movie, and mathematic modeling that accompany it, show how the rockets will disrupt the landing site (and how radioactive dust might spread if aliens attack a nuclear power plant).

On sol 3, Nilton approached Bob Bonitz to ask if the robot arm camera (RAC) could see under the lander. Nilton wanted to see if his predictions were correct.

"Why not?" Bob replied. The image was Snow Queen, the same one Nilton proclaimed his favorite last week.

Snow Queen was an instant hit. It showed the smooth and reflective Martian material.

"We were right!" Nilton thought when he first saw the Snow Queen image. "I called Manish first thing. We got some emails from colleagues who said we were wrong. But it was exactly what we thought. We could see the ice."

Snow Queen was featured in an early press conference. There was no mention of Nilton or his lab pre-launch experiments. The University of Michigan wanted to do a press release with the movie before the mission, but the mission did not approve it.

"That was disappointing," he says. "It's hard to trust some members of the media team."

Now Nilton spends a lot of time looking under the lander for hidden gems.

PETER AND DARA STAND AT THE BACK OF DOWNLINK, WAITING FOR kickoff to begin. Dara is sticking around the SOC to consult on the next sample delivery.

"Will you ever do it again?" Dara asks Peter. Peter looks a little surprised by the question.

"Not soon," Peter says after hesitating for a minute. He laughs uncomfortably.

"It takes a lot out of you," Peter says with a sigh. "I can't be at the front of the room. I have to make sure they get funding. But I have full faith in these guys. I don't want to do their jobs for them. They know better than I." Peter sounds a little beaten when he talks to Dara. Maybe it's the stress of getting funding for an extended mission after the initial ninety days. Apparently, that's not going well. Chris Shinohara told me the grants they'd hoped for to keep people employed at the SOC doing data analysis after the mission aren't looking promising. It's weighing on Peter today.

It's a rush to get on the surface and working, but now there's only two months to figure out how to keep everyone working after the mission. Dara isn't sure about the future.

"I'll be headed far away from email and computers," he says.

Bob Bonitz, Doug Ming, and Mark Lemmon look at a 3D image of the new trench. They closely examine the unexpected dark stuff.

"I think we need to keep monitoring until we know what's going on," Doug says. "We need a surface sample anyway, so we could grab one just to the left and deliver to the OM [optical microscope]."

"That sounds like the right idea," Peter says, joining the discussion. "It's better we understand the site before we jump in with more digging and disturb things."

A group of geologists and camera engineers congregates in the SSI office. They look at a new image: the two white chunks from Tom's talk. In this new image, taken at the same angle as the old one, there are no small chunks. They've disappeared.

"We were expecting that," an unenthusiastic scientist from JPL says over a bite of french fries.

"Well, if *expecting it* doesn't mean it's exciting, I don't know what to say," Mark Lemmon responds.

"Yes, it's very exciting," the JPL scientist says.

According to nearly everyone, this means one thing: it's ice. It's enough to convince Tom Pike. He's ready to celebrate. Everyone else hedged for so long, afraid to make any big claims, I think now they're afraid to celebrate.

"If we all agree this is ice, we should change the sol plan," Carol Stoker says. "Let's stop digging in Wonderland and get this Dodo-Goldilocks ice into TEGA." She thinks they can easily dislodge a few more chunks and drop them into TEGA. No one else seems too keen on the plan.

An ice sample is a huge win for the mission. Even NASA thought that getting ice would be the hardest part of the mission. NASA did not require Phoenix to get an ice sample to claim success. You can't require the team to discover something. That sorta defies all logic.

"This might be paydirt and we're digging elsewhere," Carol says to Peter. Carol objects to moving on, when what we want is right in front of us. She thinks that any ice sample is a great ice sample. Moving to a new digging area would waste valuable time.

"It's a mistake to start a new trench when we have what we came for right in front of us," she says.

Carol Stoker is the co-investigator who heads the biopotential group. She was one of the scientists who first pressed Peter to pitch this mission—and this was before Bill Boynton's hydrogen-at-north-pole-of-Mars discovery. She pushed him to go for it. Carol, like Nilton, is one of the few interdisciplinary scientists who thinks about the big picture. Carol framed the requirements for one of Phoenix's main goals: determine the habitability potential of Mars. That might be the coolest job on the mission. Carol is a researcher at Ames Research Center, a NASA satellite in northern California. Carol, like a lot of her colleagues here, is outspoken and doesn't mind being in the minority opinion. Carol is not attached to any instrument. Her role is to make sure that the research done with the instruments speaks to the larger goals of the mission. She won't let this rest and makes her case to Peter. She doesn't care that Wonderland is the most promising area. There is ice in front of them.

"I'm not sure I agree with you," Peter says to Carol. The rest of the group is quiet.

People begin to crowd into the SSI office. Everyone wants to see. With the crowd comes more celebrating. The influx pushes me closer to the front and I get a glimpse of the monitor. This time it's clear. Now there's chunks. Now there's not.

"It would be hard to argue that's anything but sublimation," someone says. This is the moment of discovery. Drink it in. It goes down smooth.

"Are we ready to go public?" Tom Pike asks.

"Sure," Peter says, "but there's no press conference today."

A FEW HOURS LATER, A PRESS RELEASE GOES OUT: *BRIGHT CHUNKS AT Phoenix Lander's Mars Site Must Have Been Ice.*

"It must be ice," said Phoenix Principal Investigator Peter Smith of The University of Arizona, Tucson.

"These little clumps completely disappearing over the course of a few days, that is perfect evidence that it's ice," Smith added. "There was some question about whether the bright material was salt. Salt can't do that." Salt does not melt like ice.

At the end of sol science meeting, Tom Pike does a follow-up presentation, appending the disappearing-nuggets image to the end of his controversial talk. He's very civil.

"I think we got what we came for," he says in conclusion.

"NBC NIGHTLY NEWS IS COMING," ONE OF THE SSI CAMERA ENGINEERS says. The next morning, Phoenix is the top national news story. Sara Hammond and her team schedule a press conference. "Phoenix finds ice" is the big Internet story from Gizmodo.com to MSN; it's even number one on everyone's favorite news aggregator/procrastination mecca, reddit.com.

There's a local news crew setting up an interview when I get into the SOC. One of the mission managers, Dave Spencer, says how excited he is about the findings. The NBC cameraman frames the shot with me in the background. I'm on TV typing this very sentence . . . and that makes me happy. Today feels like success.

On Mars, Phoenix digs into Rosy Red to collect a sample for the next TEGA delivery. We won't know the results until later.

"Please cross your fingers," Ray says. "Today, we de-lump and move ahead." By this time next week, we might know what that mysterious dark material is. Great discovery is imminent.

PART II

RED PLANET
BLUES

DATE: JUNE 26, 2008

"**P**HOENIX POISED TO DELIVER SAMPLE FOR WET CHEMISTRY," reads the June 23rd press release. MECA works and it's about to do its first experiment. Mike, Sam, Suzanne, and the rest of the MECA team are poised for amazing results. The follow-up press release on June 25th reports on MECA's progress, "Phoenix Mars Lander Puts Soil in Chemistry Lab, Team Discusses Next Steps." The MECA team gets their moment to shine. Buzzing in and out of the MECA office, they analyze and discuss how they might untangle the strange Martian chemistry. What new and amazing things have they found? They're not saying just yet.

The happy headlines hide an issue. Where's TEGA? Weren't we on the path to another delivery just a few sols ago? These scant and lumpy press releases makes it tough for the average Phoenix fan to know what's happening in the mission. Buried in the second-to-last paragraph, there are a few troubling lines. Something's wrong with our beloved TEGA.

"Four days of vibration eventually succeeded at getting the soil through the screen. However, engineers believe the use of a motor to create the vibration may also have caused a short circuit in wiring near that oven. Concern about triggering other short circuits has prompted the Phoenix team to be cautious about the use of other TEGA cells," the press release says.

Other short circuits? Cautious? That sounds like a big deal. "Concern" over using the other TEGA cells should cause some worry. The official updates come in fits and starts. The mission schedules a teleconference to clarify things. We've finished nearly a third of the Phoenix mission. There's lots to talk about. Let's dial in.

"Thank you very much. Hello, everybody," Jane Platt, the head of PR for JPL, says. "Welcome to a Phoenix Mars Lander media telecom for Thursday, June 26th. . . . And, today, we have some really intriguing science results that we'll get to in just a moment when we switch to the scientists at the University of Arizona in Tucson. I do want to also let you know that we have Phoenix project manager Barry Goldstein of JPL here with us today. And, he'll be available to answer any engineering questions about the mission. We're going to switch right now to the scientists at the University of Arizona. And Sara Hammond is going to introduce them. Sara?"

"Hello from Tucson, everyone," Sara says. She introduces Bill but she uses the more formal William Boynton and spells his name for the reporters B-O-Y-N-T-O-N. And says he's co-investigator and lead for the Evolved Gas Analyzer from the University of Arizona. She introduces Sam Kounaves, the wet chemistry lab lead from Tufts University; then Mike Hecht, the lead for the MECA instrument, the Microscopy, Electrochemistry and Connectivity Analyzer from the Jet Propulsion Laboratory. And finally Leslie Tamppari, the Phoenix project scientist from the Jet Propulsion Laboratory. Bill goes first.

"Okay. Thank you, Sara," Bill says. "The data coming out of the instrument is just spectacular," he says. Bill describes how they're seeing carbon dioxide and water vapor but no ice in the first sample they analyzed. This is expected, since the sample sat around for a while and any ice would sublimate—turn to vapor—before it had a chance to get into TEGA.

Bill says it'll be a few weeks before they can say anything more, but that there are lots of interesting things.

"What we can say now is that this soil clearly has interacted with water in the past," Bill says. Bill finishes his short statement without any mention of the TEGA issues in the press release.

Mike and Sam describe MECA's first chemistry results. With all kinds of caveats and "This is very preliminary" type statements, they describe what they found. The first soluble chemistry experiments on Mars are a big success.

"Over time, I've come to the conclusion that the amazing thing about Mars is not that it's an alien world, but that in many aspects, like mineralogy, it's very much like Earth," Sam says. Then in a humorous and potentially career-defining move, Sam tells the press that the Martian regolith is similar to what you might find in your back yard.

"You might be able to grow asparagus very well, but probably not strawberries," he says, as the Martian soil has lots of nutrient-rich minerals and is slightly alkaline. Everyone loves this. Even the conspiracy-theory bloggers hail it as a great quote. The next day's *Times of London* headline reads, "Ground Control to Farmer Tom." Sam will now be known as the asparagus guy. I wonder if he has regrets.

Leslie explains there are two trenches so far. The focus of the mission will turn to the second trench in the Wonderland area, "which is in a polygon center," she says. Leslie gives a brief summation of the samples and the work that been done so far.

"To date, in our first 30 sols, we've not only accomplished these first samples. But we've also completed 55 percent of our true color panorama of the landing site." That's more than nine hundred images. And there are a lot more to come.

Of course, they also gathered lots of atmospheric data.

"We take data almost every day. So we're learning a lot about the atmospheric science."

Thanks, Leslie.

There's a little more summarizing; then Jane Platt opens the phone lines to questions.

"Hi, thanks. Can you hear me this time okay?" Craig Covault from *Aviation Week* asks.

Yes, we can.

"All right. This is for Bill Boynton and perhaps Barry as well there at JPL," he says.

Okay, go ahead.

"Yesterday's release discussed additional factors relative to the doors on TEGA. And I'd like you to take just a moment, to discuss that a little bit and whether there's—I got the impression there was more than one door involved—possibly involved. It's the same kind of problem. And, if so, how did you manage to launch with that kind of latent situation?" he asks.

Craig wants to know why the doors aren't opening all the way. But what I want to know about is the electrical short. Bill says he's going to let Barry handle the questions.

"Okay. Hi, Craig. How are you?" Barry asks. Barry and Craig are on a first-name basis. Craig is a veteran reporter; one of the most respected according to his colleagues at *Aviation Week & Space Technology*. Av Week, as insiders call it, is an important aerospace trade publication—more important than *Engineer Aficionado* or *Vogue Avionics*. The Phoenix team was proud as peacocks—err—phoenixes to get their mission on the cover after landing. It was a very tasteful full-color spread.

"Okay. Yeah, let me clarify this a little bit," Barry says. "We have been working—ever since the first door opening—to try to find out what the problem was. If you recall, the first cell that we opened had one side open perfectly, and the other side was hung up at approximately 30 degrees.

"When we opened door number five, the next one, we noticed that the phenomena of one of the doors was actually on both of the doors on door number five. So recent investigation has shown that there's a mechanical interference that affects the inner doors of the TEGA. So if you think of the four cells on either side, there is one door on each side. Each of the doors on the end will be fine. The other doors will be hung up at approximately 30 degrees." This is technical language for "the doors are stuck." The protective door that covers TEGA was supposed roll back like an old-timey sardine can once Phoenix landed safely on the ground. This protective barrier would retract and the TEGA doors would be free to open. But the last roll didn't wind back all the way and there's a little piece of metal that's preventing the TEGA from opening all the way. And it's going to be much harder to put dirt in the ovens.

Not to worry. Barry says the RA team developed a technique to deliver to the partially opened doors and they worked through the problem.

"And, a quick follow-up there," Craig says. "How actually did you launch with a situation like this, that was not discovered on the ground?"

"Craig, we're still looking into that," Barry says. To some of the engineers who sit a bit lower down on great engineering totem, there's no need to look any further: a machining error caused the problem. It's a kind of embarrassing oops, a consequence of limited time and money. It was another design flaw inherited from the Polar Lander. They tried to fix the problem, but the contractor used the wrong blueprint to fabricate a steel guide rail that opens the TEGA doors. It ended up being one-hundredth of an inch too long. When it happened, they discovered the problem and sent the part to be fixed. Unfortunately, the shop used the same faulty blueprint to re-machine the part. And since TEGA was already behind schedule—it was the last instrument to be bolted on the lander—there was a risk of missing the delivery. Because of the time crunch, no one measured it before it was attached. It looked fine on visual inspection but now that it's been retracted on Mars, there's just a tiny scoonch, one-hundredth an inch, blocking the port doors from opening. It's a tough story to explain to the press. No one wants to be responsible for a mistake like that. Still it's a bit annoying that they don't address it head-on. Face the risk. But I digress. We want to know about the short first and foremost.

Bruce Moomaw from *Astronomy Magazine* asks the next question.

"On last night's release, there was a reference to some problems associated with the vibrating screen on oven number five," he finally says. Could you give us any more details on that?"

"What we found was when we opened the doors to cell number five," Barry says, "the current draw was indicating that we probably had a short circuit—generated by heat. So that was the theory that the team went on to investigate. And we ran a diagnostic check on the vehicle that fairly certainly can, uh, confirm that."

There are few follow-ups and Barry describes the short discovery in a bit more detail and how they are on top of it. So not to worry. That's not too enlightening. But for the moment, I guess it puts our fragile minds at ease regarding the future of the mission.

Then we shift away from the engineering issues. A reporter asks Bill if they're "definitely excluding organics?" He'd like him to confirm that

"You did not find any organic material and any carbonates?" I think it's the same negative guy from the "No water on the first try" episode back in those heady drug-fueled days early in the mission.

"Actually at this point, we can't really either include or exclude organics," Bill says. There's still a lot of work to do before then. But stay tuned because they'll get there. The first third of the mission is over, and we're on our way to discovering more great things.

CHAPTER TEN

I, FOR ONE, WELCOME OUR NASA OVERLORDS

SOL 36

IT'S JULY 1ST. CARLOS LANGE STANDS ON A DESK AND TACKS A Canadian flag to the atmospheric science theme group's work space.

"It's Canada day!" he says in his Brazilian accent. Carlos selected a sol quote:

> God bless America, but God help
> Canada put up with them.
> —ANON

Everyone seems to be in a good mood. Data analysis and discovery keep the MECA team busy. They hide out in their little lab or sit in their office and argue. Mike Hecht and Sam Kounaves, usually quite stoic, look almost giddy when they walk through the halls. Since the press conference, they haven't said much more. They're not divulging what exactly gives them the giggles—they're a tight-lipped bunch. Still, the initial measurements of pH and a surprisingly Earth-like chemistry of the regolith puts smiles on our faces. A little oxidation-reduction is all a scientist needs to feel new.

The MECA team is not the only group making progress. There's good stuff happening everywhere. The LIDAR team on MET finally gets some data to crunch. LIDAR, if you recall, shoots a powerful laser beam into the sky—Pew! Pew! Pew!—and then measures the light scatter that bounces off the atmosphere. They use a technique similar to RADAR to chart what is happening in the arctic atmosphere. Yestersol, the Met and LIDAR did an experiment where they fired their laser at the relay orbiters to conduct simultaneous measurements of the atmosphere. The atmospheric team worked with the Mars Reconnaissance Orbiter and Odyssey satellites currently orbiting Mars in order to conduct simultaneous measurements from the ground and space. These coordinated efforts sharpen the climate modeling capabilities of the scientists. This is a key part of the mission: understand climate change on Mars. Because Mars doesn't have an ocean, which would create all kinds of complicated weather phenomena that are unpredictable, it's easier to build climate models. We already know that several million years ago Mars experienced dramatic climate change. But we don't know a lot about what things were like before that. So maybe, just maybe, understanding Mars's climate might one day—hopefully, maybe, possibly—keep our blue wet planet from looking like this red dry one.

It's time for TEGA to get back in the swing of things and shake off the sticky-soil and tented-door problems. The RA team thinks they've just acquired a nearly perfect sample for TEGA. They practically glow with pride.

The sample from the Wonderland site is in a trench just called Snow White—not to be confused with Snow Queen, which is under the lander. The sample comes from the ice-soil boundary that the geologists, chemists, and nearly everyone else believe is the most dynamic.

"When we wrote the proposal, this was the sample we had in mind," Mike Mellon says. It's the mostly likely to have organic material and other interesting mysteries that aren't as obvious from the surface samples. Mars conventional wisdom says that the harsh UV rays from the sun alter the chemistry and make everything kind of boring. At the ice-soil boundary, the dirt is insulated from these harmful effects, and can possibly yield some amazing results.

Rosy Red is the name of the sample the RA team just acquired. After a painstakingly long night of writing robot arm code, the RA team is wearing big smiles.

"It's perfect," Ashitey tells me. "Exactly what they asked for."

Vicky Hipkin takes us through a breezy kickoff. The flash-memory anomaly is nearly solved. They're making progress and they're at about 80% capacity. The RA team mastered dirt acquisition obtaining the Rosy Red sample for TEGA, and the instruments are all performing well.

"How about a weather report from Palle?" Vicky asks.

"No major changes. No storms," Palle replies.

"You're going to surprise us one of these days, Palle," Vicky says.

"I doubt it," Palle says.

"CAN EVERYONE PLEASE GET TOGETHER FOR AN ANNOUNCEMENT?" Peter asks. It's really a statement. He isn't smiling or looking too enthusiastic. He waits for the SSI, RA, RAC, and mission management offices to empty. He wants everyone to hear this.

"Mike Griffin had a discussion with Ed Weiler," Peter says. Mike Griffin is the head of NASA, appointed by the President, and Ed Weiler leads NASA's science program.

Peter pauses for a moment.

"So now we have a directive from NASA. We can't take another sample until we get a sample with ice in it. Now this probably came all the way down from [President G. W.] Bush," Peter says sarcastically. The President wants in on the Mars action? Doug Ming frowns. Mike Mellon folds his arms. Confused and angry looks ping-pong around the SOC.

Why would the head of NASA tell Peter what to do? This is a direct order to stop the mission and not proceed until they put ice in TEGA. Peter says they are not interested in debate.

The short circuit on TEGA is not a minor problem. A JPL team of experts—a special group known as a tiger team—looked at the electrical current flowing and decided the entire instrument was at risk of failing. The team determined that the next sample could be the last:

total instrument failure. A failed instrument would reflect the whole Mars program poorly in the press, and NASA administrators have made an executive decision. NASA wants Phoenix to do everything possible to ensure that this sample, possibly the last, has ice in it. That way, they can declare success even if the instrument fails.

"Headquarters says 'follow the water' is the number one strategy. So even though we have a great sample, there's probably no water in it, or at least, a very low probability. I think it's unprecedented that a head of NASA would tell a P.I. which sample he should take. I was going to resign over this. But I can't give up on everyone that easily," Peter says.

The SOC audience starts to grumble. I thought we'd just proved there was ice last week. Why do we need to do it again?

"We did discover ice, lots of people have discovered ice," Bill Boynton tells me, making it all the more confusing. The images taken with the filters of the SSI camera offer one kind of proof. The scientists use these filtered images to measure the spectral properties of the material they're looking at. This means that the light coming off the material indicates that it's ice. Still, that's not enough. They brought along their TEGA and they want ice inside TEGA. Using all the fun hardware they brought is how you prove it's all working like you expected.

This is not good science. But they are bureaucrats who think this is for the best. So much for my objective reporting. Peter quiets the grumblers and continues.

"Now, of course, we are on ice and [Mike] Mellon predicted we would find ice at 10 cm. Now we've shown it's here and at 5 cm," Peter says. "Still that's not good enough for—"

"We have the scoopful we came for on this mission, and we should not waste it," Mark Lemmon, the co-investigator for the SSI and Mars imaging master, says, interrupting Peter. Mark repeats the oft-heard sentiment that there is no better sample than the one currently sitting in the scoop. Yet, somehow Washington disagrees. Doug Ming makes an audible sigh and kicks his boot into the carpet. He looks pretty pissed-off.

"Headquarters suggested that we could dump it on TEGA and not turn on the solenoid, saving the sample," Peter says.

Wait a minute.

Peter was going to resign? This kind of non-science decision-making is exactly what Dara Sabahi had warned me about. This is not a scientific strategy. Doesn't anyone care about the scientific method? The Phoenix mission is part of the Scout program. Scouts are innovative P.I.-led missions. Unless I'm mistaken, NASA just crapped all over that idea.

Peter has brought us this far. He should take us the distance as he sees fit. There are angry red faces and jeers.

"That's all for now," Peter says. He will conduct a special meeting later today to determine what's next.

WITHOUT A CLEAR DIRECTION ON WHAT TO DO WITH THE SAMPLE IN THE scoop or how to proceed with digging, most of the activities in today's plan focus on atmospheric data and several coordinated observations that were planned a while back. Midpoint meeting starts late.

The SSI team has several graduate students who train along with the engineers to assist with basic camera pointing and image production. They collect data, fill out reports, and provide updates to the team at kickoff and midpoint. Today's SSI IDE is a young, relatively new U of A student.

"We didn't get any data," he says casually. What he means to say is they're still waiting for images. But that's not what he says. He said he didn't get data. Period. And today is not a good day to run afoul of process. An easy mistake for a budding space scientist, but a JPL land mine.

Julia Bell, a slight woman with glasses and shoulder-length hair, stops the meeting.

"Excuse me?" She asks with a pregnant pause.

Julia is today's mission manager. That makes her the captain of the engineering team. Her job is to get data uploaded to the spacecraft, and she is responsible for the second shift getting the information it needs from the first shift. She carries a huge 17" silver laptop that covers most of her tiny frame, giving the impression she's a poorly drawn 1950s robot. But when she interrupts, the room falls quiet.

"No," Julia says. "You can't say 'you didn't get data.' Because you *did* get data! You got *channelized* data! And how do we know your instrument wasn't turned on accidentally or that there is another issue?" She snaps. There is no response from the young scientist or anyone else. Maybe she's upset over NASA's imposition. She copes by tightening ranks and making sure discipline reigns during this difficult time. Her approach will help prevent mistakes, but her path to perfection might leave a wide swath of battered and bruised engineers in its wake.

"The MET and RA did not report channelized data properly, either," Julia says. Channelized data is the engineering data that tells you about the health and safety of your instrument. So if you ever find yourself in Mission Control, and you don't get any pretty pictures downloaded, but you still get a little note from Phoenix saying your instrument is healthy, don't ever say: "It's nothing." It's not nothing to the spacecraft engineers who rely on it to make sure the science team makes grand discoveries. Rather, it's critical for the systems engineers' understanding of the lander's status.

Not properly reporting your data is a kind of sin perpetrated against JPL process and the lander. It shows a profound lack of respect for protocol. And when you sin against the lander, you sin against Julia Bell. Why does Julia get so upset when you mess with the lander?

One of the secrets of Phoenix is that it wasn't actually born a lander. Well, more precisely, its body was a lander but its internal software started life as an orbiter. Julia engineered the lander reassignment surgery—a robot sex-change operation. That makes her more of a surrogate mother/reconstructive surgeon than engineer to Phoenix. Since the lander wasn't originally designed to coordinate complicated activities, she had to invent a system to enable it. Julia created the core architecture for how Phoenix implements activities and shares data between its instruments. Completing this reassignment surgery required years of focus. Rumor has it that she did it without any vacation or time off. Never.

"We need her," Peter says. "If something happened to her or she got hit by a car, we probably would have to cancel the mission."

When Julia finishes the verbal mauling of the SSI engineer-in-training, there's an awkward silence. Poor guy has bits of shredded ego all over the place.

Now that graduate students who never went through Phoenix training are filling important jobs, there's some justification in cracking the whip to keep the younguns from making youthful indiscretions. When her protocol is broken, she's not afraid to let you know. Except for her impossibly high standards, she's usually very kind.

It's a big operation to turn an orbiter into a lander. Orbiters go round and round planets. You probably knew that. Julia designed a system of interoperation that lets the orbiter software act like a lander. That made it possible for the computer to use a series of sequencing engines to turn time-based operations that an orbiter uses into logic-based operations that a lander needs so all the instruments can coordinate their activities.

"TEGA is on the deck. And might just fling dirt at other instruments to get some attention," Dave Hamara says, doing the work of a rodeo clown and running interference to save the poor young SSI engineer from any last vestiges of Julia's wrath. It works. He gets the meeting moving again.

Cameron Dickinson, the SPI I, takes Dave's cue and starts to talk through the plan.

"Then at 11:00 there's a TECP wind/humidity activity with no RA move—" he says.

"Wrong! You're giving too much detail," Julia says.

We don't get very far. This is a teachable moment.

"I'm just starting my shift. I need to know what has happened and what I can expect for the rest of the day. That's all." Julia says. She continues to describe her role, his role, and everyone else's roles.

The engineers bow their heads in shame. Eventually Julia's rant ends and we're all wiser about how the SPI I should communicate the sol's proposed operation to the shift II engineers and mission managers.

It's been a long lecture. When I look up, I see something kind of cool. It's 6:32 p.m. local Tucson time and 6:25 p.m. on Mars. Our Earth and Mars clocks have just passed one another, like two ships in the night.

This means we just went through one full cycle of the Mars clock. 36 days of adding 40 minutes to keep up with the Mars rotation. That's a full 24 hours gone. Now we've lost an entire Earth day and the clocks are almost back in sync.

SHAKING IS THE CULPRIT. REMEMBER THE INITIAL PROBLEMS OF MARS dirt being too sticky? Now it's back to haunt us. Back then, the team was desperate. They were willing to do anything to get the dirt in TEGA and salvage the mission. So they shook like mad to get the dirt into TEGA. But the thing about babies, and sensitive scientific equipment, is that you shouldn't shake them. The solenoid that does the shaking was tested at 30-second or one-minute intervals. Nobody tested the solenoid for 50-minute runs. At the time it didn't matter. They just needed dirt in TEGA.

Even though it seemed like a minor problem, all but solved when Barry discussed it at the press conference, it's not. A team of crack engineers analyzed the electricity flowing through TEGA and found that a few of the TAs (ovens) now used twice as much electrical current as they should. After punching numbers into a giant calculator, pausing only to push up their glasses as they slipped down their deeply concentrated, sweat-soaked faces, they traced the problem back to a short in TA-4.

Way back on sol 4 there was another short in TEGA. That short was a wire that had probably snapped on landing but somehow un-shorted itself over the course of the next few sols. That short was not critical because it was on the "low side." Which means it was happening at a terminal branch in the electronics. So if it went, it wouldn't take anything else with it—sort of like the last bulb on the Christmas light strand. This new short is a "high-side" short. So, if this goes, all the downstream electricity would be lost—the first bulb in the strand. And that's the end of TEGA.

No one wants to say who is at fault. So I go ask Nilton. He says JPL failed to confront the issue when problems first arose.

"It was poorly managed," Nilton says. "It ended up being one of the most expensive instruments on the mission."

"I'm not sure how Nilton knows that. He was in Michigan while we were in Tucson building this instrument," Peter says. "TEGA had its share of problems, but I don't believe it was the most expensive instrument on the lander."

It's a complicated story, because space instruments are hard to build. TEGA was a legacy.

"These instruments are very complicated," Bill says. I believe him, since I wouldn't know the first place to begin if I had to build an atom-weighing machine that worked on Mars.

"It's easy to build in a lab but when you have to miniaturize and space-harden something that sensitive—" Bill pauses, leaving me to make my own conclusion.

They actually made plans to bolt it on the lander at Kennedy Space Center if things got really hairy. Luckily it didn't come to that. Still, the wiring issues were a big distraction and probably contributed to the problems with the TEGA doors. Problems plagued the small, overtaxed TEGA team.

AT MIDPOINT THERE'S A SIGN ON THE CONFERENCE TABLE. "TEGA ISE for Hire (Cheap)." Dave Hamara put it there. Poor TEGA.

Midpoint starts where kickoff left off: Julia Bell points out yet another problem with reporting protocol. This time it's with the LIDAR. I think she takes issue with how they're describing the EVRs (Event Reports). These can be either errors or problems or sometimes just events Phoenix likes to talk about. Jim Chase insists that what she thinks is an issue, is not.

"The problem is well understood and documented by the MET team before the mission," he says. It's got something to do with the operating temperature of the LIDAR. He's really making a good argument. Julia seems to take it all in. Satisfied with Jim's assessment, she begins to back down. In his closing remarks he says: "The LIDAR is *safe*." Oh, no.

When Jim Chase says the last word of that sentence . . . safe, you can almost see his head snap forward to try to swallow the words before they reach Julia Bell's ears. It's too bad they don't use instant replay in Mission Control.

Ordinarily, it might not be a big deal. Not today. Julia Bell is dead set on restoring discipline to our rag-tag group of partisans. What Jim means is that the LIDAR is not in any danger. That's not what he said. In Mission Control speak, there are some words that don't mean what you think they mean. This is one. Safe is not always safe.

Julia unleashes fury from a five-foot frame. Hellfire and brimstone rain down upon poor Jim Chase. Hardly seems like a fair fight. While Jim could be easily be the center for the College of Engineering basketball team, Julia's build is more diminutive. Still, Jim doesn't stand a chance.

"Safe" is a safe word in Mission Control. In the SOC, the safe word is "safe." Whether it's green balloons, watermelon, or Humpty Dumpty, most folks choose a safe word that they aren't likely to say during the heat of the moment. Safe does not mean everything is okay. *Safe* means the instrument has shut down to protect itself from danger. If your instrument is safe or has safed, you must, please, look to see what's gone wrong. All the instruments can go into safe mode and they do it often. It's doesn't mean they're broken—although it could happen—it just means they're outside their comfort zone.

"I want an ISA to document this," Julia barks. Poor Jim. Now he'll spend his evening writing an Incident Surprise Anomaly (ISA). He picked the wrong word, now he pays: in paperwork.

EVERYONE MARCHES TO THE LARGE CONFERENCE ROOM. PEOPLE LEAN against the walls and sit on the floor. Peter doesn't bother to stand at the front of the room.

"Now is the time to voice your opinion," Peter says from his seat. "The new directive is to rasp an icy sample and get it into TEGA. We must treat the next cell as the last one," Peter says. NASA is essentially freezing the mission—no pun intended—until they can get a scoop of ice into TEGA and turn it on for what they believe will be the last time. And hopefully, it won't blow the whole spacecraft to bits.

Just to make things more tense, Peter agreed at the outset of the mission to give everyone the 4th of July weekend off. So they really only have tonight to sort out a strategy and then one day to implement before everyone leaves for two days.

Bill Boynton is particularly annoyed.

"This will not be the last TEGA operation," he mutters. He's looked at the data and he doesn't share the same concern as the head of NASA.

"So that's not the sample in the scoop?" Carol Stoker asks. It's a loaded, leading question. I sense a bit of passive aggression, if I might pop-psychologize for a moment. Carol wants Peter to repeat the justification and point out how ludicrous it is to throw away a great sample. She's probably pissed because no one listened to her when she suggested getting an ice sample from Dodo.

"No, that sample is unacceptable to NASA," Peter says with a sigh. He doesn't take the bait.

"Well, then, the smart thing to do is get a sample from the D-G trench [Dodo-Goldilocks], and move on. There's no point in wasting time with scraping Wonderland. We can do it later," she says.

Peter wants the team to make a scientific decision. But the science is tangled in politics; smeared with bad timing and corroded with poor TEGA wiring. One oddity that's important to note is that an ice sample is not actually part of Phoenix's level-one requirements. NASA didn't ask for this. Probably because you can't require a mission to discover something that you're not even sure is there.

Now that the process-obsessed NASA—requiring Phoenix to file ten million documents outlining its capabilities and space-worthiness— sees the opportunity for a great headline, they suddenly change their whole philosophy. As a bonus, they're undermining everything it means to be a scientist-led mission. We can discuss that later. Right now, we need to plan a mission.

There are two clear paths from which to choose. The team can take a sample with ice from Dodo-Goldilocks (D-G), or from a trench in or near Wonderland. Both have problems. Getting ice from Wonderland requires learning how to get ice. The RA team has a good idea of what it might take and they practiced with their buckets of concrete, but the real thing is different. It will take time. And no one knows if it's even possible. The RA team could spend weeks learning how to acquire ice, only to find it's impossible to gather it in any meaningful quantity. So that's a strike against the Wonderland site, specifically the Snow White trench. The problem with D-G is that the geology cabal doesn't like the ice inside it. These are the hardliners, and they just took a stand against this liberal ice. They take issue with its provenance. The D-G ice is on the boundary of a polygon. The white material in D-G is likely frost that's collected over time. The ice at the center of a polygon is what they want. The trench is also at a funny angle and hard for the RA to scoop. It was just supposed to be a practice area. Wonderland is what they consider to be the interesting area.

Peter knows using the rasp at this point would be a long process. They're going to spend a lot of time getting up to speed. Committing to Wonderland means that all exploration trenching will stop while the RA team focuses on how to crack the ice.

"We don't have approved sequences, but we will soon. It's one of our highest priorities . . . maybe approved by sol 45," Peter says.

Joel Krajewski laughs.

"Or maybe 85." Peter corrects himself facetiously.

Joel's team needs to develop a lot of code blocks if they want to even attempt an ice scrape.

"There have been eight new top priorities in the last eight days," Krajewski says, implying the science team needs to pick a direction and stick with it.

"Is that a complaint?" Peter asks, a little perturbed. He's had a hard day and is running a little low on patience.

"I'm just trying to help," Joel replies.

The process for coding new activities is slow. Ideally a mission would land with a whole library of activities coded and ready to go. Phoenix didn't have the money to do that. The engineers put their effort into a safe landing instead of surface operations.

"It wouldn't have done much good to arrive with every block validated and no lander," Joel told me early in the mission. So the engineers code and test the activities in the library as needed.

"Rasping would be two to three weeks before it could produce a sample we are comfortable with," Peter estimates. He suggests if they're going to go the Wonderland route, they should try with just scraping.

"We've never sampled small particles," Carol Stoker says. She's going to make her move. "We were going to [sample] in Dodo, but we stopped that path. The block is written and we could do it. One day! And done," she says. Had the team listened to her before, this wouldn't be an issue. She makes a passionate plea: going after ice in Wonderland is a wild goose chase.

"Before we know it, it'll be sol 90 and the mission will be over," she says.

Looking for some support, she gets scientists averting their gaze instead.

"I would argue against using Dodo," Bill Boynton says in a sort of apologetic way. "If we get that [D-G] sample, we'll find that it is ice . . . and that's it." That means that there's little discovery value in taking this simple approach.

"So scientifically speaking, Snow White [Wonderland] is the best bet for getting organics. I would further argue for scraping. We understand it better and can do it sooner. And we have had several samples that would remove any Earth organics; but the rasp has not been used as much, so it might be contaminated."

"That's a good point," Peter says.

"Well, what's the difference between what's in the scoop in this sample? It will be the same and HQ doesn't want that," Carol says.

"Our ice has sublimated!" Peter says emphatically. Probably because there's no point in that line of argument. There's no way to convince NASA that it's a good sample. Yes, we all agree it's the best sample. It's the only thing the science team agrees on. NASA doesn't care.

"She's mixing up the issues with the facts," Rich from the RA team says to me.

"I agree with Bill's statement," Doug Ming says. "And would say we need to show that water ice is in the scoop." He makes the distinction "water ice," because technically it could be some other kind of ice, but don't let that confuse you. He's very particular.

"We put a rasp on the scoop for this very reason," Mike Mellon interjects. "We didn't think we could get enough material [by scraping]." There's a lot of back-and-forth. They discuss the best methods of scraping and rasping. Timing, verifying blocks and every element is discussed ad nauseam. What's not discussed is going back to Dodo.

Carol won't go down without a fight.

"We already know there's ice in D-G. They don't know what's in Wonderland. It's a risk," she says. Putting ice in the instrument is scientifically new and satisfies NASA's requirement. The trenches are different parts of the polygon shapes everyone was so excited about back at landing. Wonderland is the center of the trench. D-G is at the edge in a boundary trough. The ice there would be younger and possible blown there from other parts of Mars.

"I would agree with Carol," Nilton Renno says quietly.

"I think we should come up with what we want and then convince HQ it's what they want," a scientist from the geology group says.

"Exactly," Peter agrees.

"If we scrape again [in Wonderland] and we generate more material, we're home free," Bill says, shifting the conversation back to Wonderland. He thinks they won't learn anything new from D-G.

Half the team puts a premium on ice that's of greater scientific value. The other half? Well, Carol and Nilton think all ice is scientifically interesting and think D-G is as good a place as any. Flip a coin?

There are a lot of loud voices in support of Wonderland. Carol and her sole ally, Nilton, can't muster any more support.

"Carol and Nilton would argue we do have H_2O signature in the other trench," Peter says to the vocal group of Wonderlanders. He doesn't want their insights lost because the other opinions are expressed at volume.

That just encourages more loud objections. Peter manages to quiet the team.

"If we just want to give HQ a happy feeling about ice, we should go for Dodo; but it's not obvious to me that if we get that sample, it will answer any of the questions about how it got there," Bill says.

"What's the difference?" Mike Hecht asks,

"Snow White is vapor deposited," Carol says.

"There's no evidence that one was a liquid or not," Mike Mellon responds. They start to fight about the source of ice. Things get real technical real fast.

"Let's hear from some others," Peter says. "Joel?" Peter knows that whatever they decide, he wants everyone to feel like they've said their share and made their peace. After this morning, all bets are off for the mission plan. It's really terra incognita.

"Anyone on the phone have an opinion?" Peter asks with a bit of desperation in his voice.

"Ben Clark, here," comes a weak voice over the phone.

Ben Clark, the elder statesman of the mission, calls in to most meetings and has a shadow presence here.

"Well, left or right?" Peter asks.

Left is D-G. Right is Wonderland. Clark is not quite ready to make a choice.

"On Snow White we have the opportunity to get more data from [MECA's] WCL and TECP on this site," Ben says. Snow White is the site in Wonderland where they're going to get the sample.

"This is the center of a polygon," Mark Lemmon interrupts. He points at Wonderland. "This is what we wrote our proposal for. That's what we came here for," he says.

"Yes, the interior is what we want. The cracks are incidental," Bill says in agreement.

"But we could spend a month trying to get that sample. I have a one-day operation that could show us what we want," Carol makes a last desperate plea.

"I don't think you can assert that it will take a month," Mark says dismissively.

"Let's hear from the RA team. Rich?" Peter says.

Rich Volpe, the RA engineer on duty, thinks for a minute. I suspect he's hesitant to opine in the science meetings. The RA team fancies itself as Switzerland in these debates. They never express their preference unless it's a technical matter. The RA team treats the scientists as clients.

"We expect to see the same rate of accumulation [in both trenches]," Rich says.

"Did you see the scoop getting lower?" Peter asks. If they can see that the RA moved lower to dig once it hit hard stuff, that might be evidence that they are accumulating when they scrape rather than just pushing around the dirt that falls back into the trench.

"We could review the telemetry," Rich says. That means he doesn't know.

"We need to design a test of scraping to find out," Peter says. All they need is enough icy material for TEGA to register it. It would look like a little peak on a graph right around zero degrees Celsius.

"It doesn't have to be a lot. Bill has a very sensitive instrument," Peter says.

"I'm a sensitive guy," Bill says.

"I'll put that in the daily report to HQ," Peter says. Peter thinks the popcorn has popped on this debate. It's been going on for almost two hours. He takes a poll.

Wonderland wins a majority. Carol and Nilton lose their case.

"I think we have a plan," Peter says. "Nilton, don't look so unhappy."

"I'm not unhappy," Nilton says.

"If you really want to be unhappy, join TEGA," Bill says.

I start to think about why it's so important for HQ to have further proof. I conclude that something fishy is going on despite my lack of evidence.

"There is a lot of work to do, to plan and execute this," Joel says.

"We should first send a plan and make sure HQ doesn't freak out," Peter says.

"I think we need a plan and decision tree starting tomorrow," Bill says.

Carol is not happy.

"We've declared every sample precious. Are we going to do something with the sample we have?" She makes one more plea to at least use the sample that's still sitting in the RA scoop.

Leslie Tamppari, the project scientist, assures her that, yes, they will deliver it . . . somewhere.

BEFORE ANYTHING CHANGES, THE PLAN IS AS FOLLOWS: DURING SOL 38, deliver a small sample we have in the scoop now to the optical microscope. Then on sol 41, deliver the rest of sample to the wet chemistry lab (WCL). Then prepare the trench for an ice acquisition. Great it's settled. I'm feeling proud that we hung in there with this terribly important decision in the mission. So proud, I consider calling it a sol.

"How can we be sure the sample will have ice in it?" a voice from the back of the room asks.

Everyone sits back down. A new debate ensues.

"That may be a catch-22," Dick Morris says. Morris describes how documenting ice would actually prevent its successful delivery. Since Phoenix doesn't operate in real time, proving it's ice requires taking an image of what you scoop up. By the time you would see the image, a lot of time would pass. Then you'd have to implement the next part of the plan. More time passes. It would be the next sol and the ice is gone—sublimated away.

"Hmm . . . we can't really satisfy both those requirements," Bill says with a chuckle.

Thinking for a moment, he realizes the whole idea is ludicrous. "If we could say it's ice, we wouldn't have to put it into TEGA," Bill says. "NASA's request is logically flawed." They're basically saying you have to put ice in TEGA so we can prove there's ice in TEGA. Apparently NASA makes a big distinction based on which tool made the discovery.

"We have to get this very hard material into the sample," Peter says, trying to clarify. He wants to steer the conversation away from distracting tautological exercises. The team murmurs. He tries to get them back on track. We just need to put aside that it doesn't actually make sense and decide on a plan, so Peter can report the plan back to NASA. Bill suggests putting the scoop on the ground as a heat sink. To try to keep the ice from sublimating, "if we're forced to take NASA's request literally."

"Yes, that would be a good way to preserve that," Peter says, happy that everyone is problem-solving instead of bitching.

Dick Morris worries that the scoop might stick to the ground.

"Like if Nilton put his tongue on a pole during a Michigan winter," Morris musses.

"This is a mess," Mike Hecht says.

"Yes," Joel agrees, sitting back down.

Mike Mellon says he thinks there's another practical problem with getting ice.

"TEGA needs its sample first thing in the morning. This complicates the delivery," he says. Using the instruments early in the morning requires heating. That hasn't been done yet. This process requires heating TEGA, RA, and RAC. The pre-heat blocks they need aren't ready. Plus, they're complicated. The extreme cold of the early morning could seize up the joints for the RA and cause all kinds of problems with the RAC images and TEGA. They could even blow a fuse if they calculate the heating times incorrectly. And if I've learned anything from putting my fork in an outlet, it's that fuses are important. There are several complicated power modeling simulations and loads of blocks to write before an early morning delivery is even possible.

Our little decision tree turns into a forest of problems each time someone chimes in with a comment. The debate spawns all kinds of offspring, and inbred mini-debates take root. Instrument teams argue amongst themselves, with other groups and systems engineers. There's an hour-long discussion of using divot images—these are close-up RAC pictures—to verify that it's ice. It's going to be another long night.

"What if you get no sample at all? We'll need another tree," Carol says.

Carol's question turns the debate back on its head, and we start over. Maybe D-G is the right choice. Oh, no. Has the core premise of what HQ is really after been met?

"If we're not qualified to make that judgment, then we're not the right people to run this mission. If they disagree with that, it's their problem," someone at the back of the room shouts. Yeah. We can't let NASA HQ decide how we operate.

This argument shines a light on why complicated processes have strict protocol. Julia Bell, however, is not present for the meeting. She's in the spacecraft room preparing to upload tomorrow's gimpy plan.

"This is not a 1 or 0, it's gray," Mark Mellon says, mixing his metaphors. The problem will not have a clean solution, but we need to move forward.

Joel steps in to help organize things. He makes a list of blocks that still require validation in order to scrape in the short term. They have to turn off some of the safety features they've built into the RA so it keeps digging in the hard stuff. It's a long list. The meeting reaches the three-and-a-half-hour mark. People fade. Heads bobble.

Finally, after about four hours, everyone is satisfied with the theoretical foundation for the plan. It's time to create an actual plan.

IN THEIR PUSH FOR "GOOD NEWS," THE EPO (EDUCATION AND PUBLIC Outreach) team just sent out a press release that says they just scooped up the perfect sample. The release reads: "The scientists saw the scrapings in Surface Stereo Imager images on Sunday, June 29, agreed they had 'almost perfect samples of the interface of ice and soil,' and commanded the robotic arm to pick up some scrapings for instrument analysis." Now it'll take some quick thinking and good spin to explain why they didn't put the sample into TEGA. I'm sure this will make the cranky *LA Times* reporter very happy.

Finally, they agree to a strategy to scrape and deliver within the next ten sols. It'll be a mess, but at least there's a clear path to scrape and deliver, get NASA off their backs, and finally discover more good stuff.

CHAPTER ELEVEN

ARM UP. STAND DOWN.

SOL 37

WE DRAG OURSELVES INTO THE SOC. MOST OF THE SCIENCE team only slept a few hours after last night's marathon session of re-planning. The all-nighters feel more and more painful. There are a lot of shell-shocked, thousand-yard stares in the faces of everyone swiping into the secure entrance for shift I.

Nothing looks different in Mission Control. There are no armed guards or metal detectors posted at the door. But don't be fooled, we are now a mission under NASA occupation.

Doug Ming is the science lead. He's eager to get kickoff started. And he starts with some good news.

"We now have 90% of mission success pan complete," he says, half-smiling. The mission success pan is a photo mosaic of the whole landing site. One of the scientists in the back, desperate for any good news, starts to clap. He's the only one and gives up quickly.

"By sol 39 it should be complete," Doug continues. He wants to lift our spirits with news of a big accomplishment. The team calls the image the "Peter Pan." But it has some official boring name—the mission success pan—that they use to keep things on the up-and-up. The image is a high-resolution full-spectrum panoramic of the landscape. It shows off Peter's SSI camera design.

"Last night, after a monumental effort, Mike Mellon created a new sol tree for our TEGA sample acquisition. Of course, we still have a great sample and we want to deliver it [to MECA]. Then we'll develop our method to deliver ice. It's going to be a bit of trial and error. For sol 38 our main activities will be the OM [Optical Microscope] delivery of the sample in the scoop and WCL [Wet Chemistry Lab] thaw. TEGA door open and SSI images," Doug reports.

Doug says there's one big caveat to the plan.

"Obviously, we have a forward dependency of HQ approving," he says. That's the only mention of NASA at the kickoff meeting.

Peter had promised the University and the Jet Propulsion Lab he would give everyone two days off for the July 4th holiday. That leaves one sol, today, to create three sol plans. Not everyone is happy about walking away from Phoenix during this crisis. Although, after more than a month of this, everyone needs a break. If we manage to plan three sols today, a skeleton crew will oversee ops at the SOC. To pay penance for missing two great discoveries earlier in the mission, I will stay with them. Everyone else will drink beer and barbecue.

"We modified the schedule," Bob Denise says. I guess that's necessary for simultaneously planning three sols. He will do activity plan approval meeting (APAM) for sols 39 and 40 right after kickoff. APAM is where the shift II team digs into the science requirements and nails down a plan. Once the plan gets APAMed, sequencing begins. This is the sausage grinder that Dara Sabahi talked about. Sequencing these simplified, almost hobbled, plans is not the usual gory mess. This should be tamer, mere hot-dog making.

The plans for sols 39 and 40 consist of remote sensing activities that already have completed sequences in the library. It's a bit of a copy-and-paste job from known activities that Phoenix can do with its eyes closed.

Then the rest of the day will proceed as usual for planning sol 38. I go to uplink to watch Denise's ultra APAM.

PETER IS ALONE IN THE KITCHEN, WARMING UP SOME SOUP. HEATHER and Chris made some communal pozole soup. They're alone this summer. They sent their five kids off to in-laws and summer camps so they could manage Mars time. Now they make large batches of soup for the team. What could be better than homemade soup and some quality time with Peter?

"It's tough to say what's right," Peter says unenthusiastically. He doesn't seem like he wants to talk about NASA, or anything really. But he gamely answers my questions, even though he'd really rather enjoy his soup in peace and quiet. That's fair. Unfortunately, I'm not getting the hint. I tell Peter that some of the team is angry that he's not pushing back harder against NASA.

"What am I supposed to do?" Peter asks, annoyed. "They hold all the cards. If I don't comply, they can withhold funds or whatever they want. Not to mention that the engineers who control the code work for JPL; they can refuse to send anything to the spacecraft if their superiors tell them to. It's impossible to move without them. We have to find a way to work together." He goes back to his soup. I try to change the subject. The few members of the team who do grumble don't really understand the position he's in. It's his mission, but without their checks, it's hard to keep his staff fed and experimenting.

"He's kept the pressure off the team and given them the freedom to do their science," Chris tells me over a beer one morning after shift.

We manage to shoot the breeze about the equipment on Phoenix. I'm just happy whenever Peter talks to me about anything, and I hope I haven't ruined his lunch. Peter gets up to have another bowl of soup. I sit there and worry that we'll never bond over peyote in the desert. There isn't going to be a half-naked Indian man taking us to find our spirit animals.

"You know, ten years ago when we built the first SSI, you couldn't buy a one-megapixel camera," Peter says. Peter describes how Viking I and II used a one-*pixel* camera and did a raster scan. They would capture one pixel. Then use a motor to move the camera. Then the next pixel. And the next. It was all mechanical, no digital elements at all.

"Now, *those* guys had some engineering challenges," he says with a slight smile.

Space imagers are not like any camera you can buy off the shelf. The CCD chips in them are carefully calibrated, and each pixel defect is carefully understood. The glass lenses must be custom-made. They take thousands of test images to "characterize" the camera because the chips make each one unique. Then calibration targets are used to understand light variations. These calibration targets are carried to Mars so that they can continually correct the images for accuracy as lighting and other conditions vary. The targets are reference points like Pantone color swatches. Some use pigment and a rubber compound. Others use crushed-up minerals that they might find on Mars, like hematite or goethite.

"The camera properties vary with time and temperature," Peter says. "These cameras need to work consistently through an extreme temperature range throughout the mission. Something between −100 and −20. That requires a careful understanding." But further explanation will have to wait. Peter finishes his soup and stands up. Okay, back to work.

ON MY WAY BACK TO DOWNLINK FROM MY PETER SMITH SOUP ACCORD, I pass Joseph Carsten from the RA team.

"You want to come test the delivery in the PIT?" Carsten asks me.

I do want!

There's a pretty severe warning on the door to the PIT. Access is extremely limited. There's even a list of names of who can enter. My palms get sweaty whenever I go through that secondary security door and walk up the ramp to the lander. I'm never quite sure it's okay. It's so damn exciting, you're pretty sure you're breaking some rule.

Joseph is here to test a hypothetical delivery plan. He'll present it to Ray and the team. If they like what they see, this will be the approach. This delivery is extra tricky because the oven doors aren't open all the way. And this particular cell sits at an awkward angle. This is the first delivery to this side of TEGA. The tented doors, the solenoid scare, and the new side mean lots of extra documentation and special care.

Joseph went to JPL after graduate school in robotics at Carnegie Mellon. His advisor's interest lay in autonomous driving software for roving missions. Joseph got his robot driver's license while he helped develop some new robot driving technology. It was great timing. The team at JPL that operated the Spirit and Opportunity rovers needed to update their driving software. They chose to implement the package

Joseph worked on. JPL recruited him to implement it. His robotics cred helped land him a spot on the Phoenix mission.

"I was a late add," he says. He moved to Phoenix from the rovers just a few months before landing.

Tonight, Joseph's working alongside Rolfe Bode in the PIT. Rolfe is the PIT test engineer on duty. I met him at launch. In fact, he's the only person who contacted me after we watched Phoenix roar off into the sky to offer me a tour of his lab. Now I feel guilty for never taking the tour.

"You want some cookies?" Rolfe asks. "My wife made them. They're chocolate chip."

Rolfe and Joseph are both pretty relaxed about the long night ahead of them. Even though they wear matching Phoenix lab coats and anti-static grounding bracelets, they couldn't look more different. Joseph is fresh-faced and lanky, maybe 100 pounds if soaked. Rolfe is the gentle veteran who looks like he's seen it all—the hearty space veteran. We all eat cookies.

Joseph fiddles with a video camera to document his practice run. His mission tonight is limited to finding the best delivery pose for this particular TA oven in TEGA. TEGA is shaped like a little house. The doors to the ovens are positioned on each side of the "roof." One side of the ovens is oriented uphill, toward the scoop head, and the other is sloped downward. This is the first delivery on the more awkward downhill side.

Rolfe says this is the first time he's worked on the surface ops phase of a mission. Usually, test engineers like Bode work just until the lander or rover launches or arrives on a planet. Then their work is done. Then the fun part comes. It's what the RA team calls the "ice cream phase." It means that the engineers who do all the work before landing have equally long hours but don't get any of the ice cream. Once the spacecraft launches, guys like Rolfe usually look for new work. Move on to the next mission, and miss out on the windfall of dessert and glorious new discoveries.

"Don't forget about those folks," Rolfe says.

There might be only 130 or so faces coming through this building, but it took a cast of thousands to get Phoenix on Mars.

"Sometimes you feel like you're not doing much to help the effort," Rolfe says. "But I guess without us, they're not going anywhere."

Joseph holds a little bucket under the scoop and they dump out an old sample. We wait. Nothing happens. We shoot the breeze. Still nothing happens. I hope it's not broken.

"Did we turn it on?" Joseph asks. He flips a switch.

Phoenix II comes to life. The arm rises and tilts its scoop head to rid itself of the old sample. Joseph is there to catch it in a little cup. Then we sit quietly while Rolfe and Joseph calculate and code.

"Let's do real-time commands" Joseph says after about fifteen minutes.

"Okay," Rolfe says, "I'm ready."

Joseph reads off the coordinates. Rolfe enters them into his machine. They remind themselves to check the sequence length. New flight rules dictate delivery be less than three minutes. They can't risk another short.

I wonder what will happen when they get the rasp going. Maybe they should have some kind of flight rules to keep it from running too long. I don't want that rasp shorting on us too.

Everything is in place. Joseph presses the red button on the camcorder. Rolfe initiates the sequence. The RA moves slowly into position . . . very . . . very . . . slowly.

Can't it go any faster? How can I get anyone excited about Mars machines if Phoenix II limps along like this? There's even a wire support hanging down from the ceiling to brace the RA joints.

"Sorry, it's built to Martian gravity specifications" Joseph says, defending his robotic arm.

"We have to simulate the gravity the arm would feel on Mars," Joseph says. Mars only has one-third of the gravity we have here, so there's a lot more pressure pushing down. Fair enough.

The scoop tilts into position, struggling and straining to cope with Earth's heavy gravitational effect and dreaming of the day it might work unencumbered on Mars. The rasp kicks on with a whirl. The vibrations reach the sample at the back of the scoop, and a narrow dirt convoy streams out. It's just like the soldiers Ashitey described. There they go, marching off the edge in an organized single file. Joseph measures the amount of sample delivered. He snaps a few photographs so he can show the team the highlights. Then they're free to choose their favorite approach.

With the arm moving at a snail's pace, each test can take an hour or more. We're going to be here a while. I don't mind at all. Testing dirt deliveries, even slow ones, is the best way to feel like you're really living on Mars.

NILTON CONFESSES TO ME THAT HE FEELS A LITTLE RESPONSIBLE FOR all this trouble with NASA.

"I sent an email to the entire team after we didn't sample the ice from D-G [Dodo-Goldilocks]," he says. The email said not dumping those little clumps of ice into TEGA was a big mistake.

"The ice could have even had salt in it," he says. "It would be a great sample. If there was briny water, the purer water would freeze out first and push the salts up and keep raising the freezing point. This is the best chance for getting salts and organics. I was really annoyed."

Nilton wrote up his concerns and emailed them out to the entire team. It was the same time the TEGA short happened. He didn't realize the short was so serious. But NASA and JPL were busy investigating the implications for the mission if TEGA failed. Some team members thought that the timing caused NASA to overreact. An angry Doug Ming confronted Nilton and suggested this note might have caused the whole mess.

There's an easy fix, in Nilton's opinion. Grab some ice from Dodo and move on. We wouldn't be here planning all night. The mission could continue like normal.

"There are a lot of geologists here that want a specific sample, and so the team pushed for the other trench scoop. But that's not the primary goal of the mission. We are here to follow the water," Nilton says. The ones who reject D-G say this white stuff is just frost deposited from the atmosphere on a cold finger and not the ice they're after. But they can't be sure unless we sample it. Nilton doesn't like the idea that they know what's best when we really don't know much of anything. They should take a step back and look at the wider goals before pushing for any one activity.

IN SPITE OF BEING COLLEAGUES AND FRIENDS, CARLOS LANGE disagrees. He says the center of the polygons is representative of the region. He thinks Dodo is the wrong choice. That's an idea that's popular with most of the geologists.

"Dodo is a boundary feature and accounts for less than 10% of the area. That's not to say it's not interesting, but all the models and hypotheses they want to test are geared for the polygon center." He makes a good argument. The team is headed in the right direction in spite of Carol and Nilton's objections.

"Then again, I'm not sure they should listen to me," Carlos says. He has strong opinions but a lot less Mars experience than Nilton and Carol, or Doug and Mike. Before Phoenix, he was interested in math models of dust moving in your lungs. Now he's obsessed with dust in the Martian atmosphere.

AS IT GETS LATE, THE MOOD LIGHTENS IN THE SOC. PEOPLE LOOK forward to an Earth day off. Over coffee, I have a mock debate—D-G or Wonderland—with Maria Banks from the outreach department. She's one of the overnight notetakers. Maria keeps the media team updated in the wee hours when no one else wants to be here and/or nothing exciting happens. It's a pretty sweet summer job. She is a graduate student at the University of Arizona, and her research requires her to stare at images of Mars all day. She's doing research on a big camera called HiRISE attached to the Mars Reconnaissance Orbiter.

We argue the merits of a quick delivery from Dodo or going for broke with Wonderland. We both agree it's not fair that NASA stepped in. Maria has some good ideas about keeping the scoop pressed to the surface so any ice would stay cool and not sublimate. I tell her Bill proposed the very same thing. Great minds . . . and all that.

Me? I'm less confident. They both seem like good options.

Joe and Rolfe are still testing in the PIT. They have hours and hours left to go. I peek my head outside; the sun is just starting to come up over the desert's horizon.

CHAPTER TWELVE

ALL THE LANDERS, INDEPENDENT

SOL 38

HAPPY JULY 4TH. SHALL I WAX POETIC ABOUT NASA GETTING TO space and the meaning of freedom? Just kidding. We're having a party! The TEGA team hosts a Phoenix party in the hills overlooking Tucson. The best thing about the party: they invite me.

I catch a ride up to a beautiful home in the hills with Peter's assistant, Frankie. The house belongs to Dr. Gerard Droege, M.D. Don't let his curly gray hair or reading glasses or medical degree fool you. He's the most junior member of the TEGA team.

"Thanks for coming," Droege greets me at the door. Chris Shinohara and Heather Enos unload several dozen pounds of meat from a giant cooler.

"There's beer over there, and the chicken should be done soon," Chris says.

"Wait! No reporters at this party!" Heather says. My stomach knots up. Doesn't she know I need a day off and beer too? I was at the SOC the whole day; I swear.

"I'm just kidding," she says. "But nobody talk to this guy!" This is hazing. Heather introduces me to Chris's brother and some other family members. I promise her everything from this point on is off the record. So I won't say anything about what happens from this moment—

THE SWIMMING, BEER, BOOZE, AND FIREWORKS PARTY IS OFF LIMITS TO any salacious gossip that would really flesh out the characters in this book. Such personal disclosures might persuade Congress to double or triple NASA's budget. But I will respect the team's off-the-record wishes.

Instead, I'll take a moment to talk about the host of the party, Dr. Droege.

Gerard Droege is the man who inhabits this beautiful house and its extremely well-curated art collection. He is a junior at the University of Arizona. And please call him Jerry.

"He's a great guy," says his advisor. She's about twenty-five years his junior and wasn't really sure why he came to the U of A to get his computer science degree. Droege has one year left to finish his degree. He's already done his medical degree, internship, and residency, and worked in private practice as an obstetrician for twenty years, but he was living a lie. He didn't love babies; he loved planets. In spite of an extremely successful practice in Westchester County, a swanky suburb just north of New York City, Jerry decided he'd had enough.

"I used to keep the *Journal of Geophysics* in the waiting room," he says. That should have been a clue that he was no ordinary obstetrician.

Every delivery, I suspect, he'd look down at the lumpy mucus-covered crying little alien heads and think, "I should be working in space."

"I always complained about wanting to change careers, especially to a particular friend. I guess he got tired of hearing it," he tells me. His friend, Leo, invited him to a party where an astronomy professor was going to give a talk.

"I was blown away," Jerry says. "I went up to talk to him [the professor] afterwards." They talked.

"There's more where that came from. You should consider taking an evening class in astronomy," said the affable professor. Jerry enrolled right away. And that was it. Professor Neil deGrasse Tyson taught the Columbia University extension course. Okay, so it wasn't just any adjunct professor, but one of the most passionate and interesting astronomers working today. Tyson is the head of the Hayden Planetarium in New York City. You might recognize him from one of the many science TV shows he's hosted. Once the seed was planted and Jerry felt his planetary baby kicking, it was obvious. He couldn't be an obstetrician anymore. All that space angst developed into something much bigger, and it was starting to crown. And just like this birthing metaphor, it really needed to come out.

"I started doing the math on how I could go back to school. What would I do with my practice? What would I tell my employees? I didn't know, but I could figure it out," he says. He could make sacrifices. And no matter how he computed, it didn't seem like any sacrifice was bigger than not following his passion.

"I was willing to start over. I was willing to go back as an undergrad and then go to grad school. It didn't matter," said Droege.

Tyson is a pretty great mentor to have, if you're going to drop your entire life and become a space man.

"He [Tyson] told me if I were serious, I should consider moving to Arizona. He said if I were persistent, I could find a home there." The next day he called his two employees, his nurse and his receptionist.

"I promise to find you both jobs," he said. "Then I told them I planned to move to Arizona and go back to school." It came as a bit of a surprise.

Then he put his practice up for sale and filled out his application.

"It wasn't that big of a deal," he says. He just abandoned his career, moved cross-country, and went to school with kids half his age.

"Well, if I failed, I could always go back to delivering babies. I was fully prepared to do four years and then on to grad school to try," he says.

When Droege arrived, he registered for classes and went looking for a job.

"I just went around and knocked on doors," he says. Droege started at the Lunar and Planetary Lab looking for his after-school job. At the time, Bill Boynton and Heather Enos were hard at work with TEGA calibrations. Overwhelmed.

"We were happy to have some extra help," Bill says. He was in luck. They needed cheap labor. Jerry wanted an opportunity. He was willing to work late, learn anything or do anything.

"He was a quick learner and more than competent. We felt lucky to have him," Bill Boynton says of Droege. Jerry worked like mad to get up to speed on the complexities of planetary mass spectroscopy. At the end of the year, he'd proved his worth. They offered him a full-time position as a technician. This was far better than anything he'd imagined. After his first year, he'd secured his place on a Mars mission. When he tells the story, all I can think about is this line in *The Alchemist*: "And, when you want something, all the universe conspires in helping you to achieve it."

"When they [Heather and Bill] told me they were going to hire me, I just kind of said, oh thank you. And that was it." They were all hanging out having a good time celebrating some achievement or the end of the year. "Then I got in my car and it just sort of hit me. I was overwhelmed. I just cried the whole way home."

Thanks for having me at the party, Jerry.

BEFORE THE PARTY, I SPENT THE DAY AT THE SOC, AS PROMISED. THERE was no way I'd miss another moment of this mission. No way. Not after the last two moments of discovery. So I sit with the skeleton crew as we go through the motions.

"You have a report for me, Pat?" Bob asks Pat Woida. Bob Denise comes into his office. It's about fifteen minutes before the final tag up with Denver. He wants the instrument reports from Pat Woida, who, along with his son, Rigel Woida, and Joe Stehly, are what's left of today's Phoenix crew.

"It's a lot easier when you only have to track down two people," Bob says. He's enjoying the calmer side of mission life.

"I've done my IDE duty for SSI, RAC, and MECA OM. They are all healthy and look good. There are a few EVRs, but no show-stoppers. They're having some compression errors on the images." Pat gives Bob the run-down, conducting what would be a two-hour meeting in thirty seconds. The plan is simplified observations and remote sensing. They're activities that have already been done. Bob takes the plan to review the errors.

"You got to see this image," Pat says to me. It's a high-resolution 3D image of the foot pad. It's a kind of science homage to the first Mars image ever taken. When Viking I landed on Mars, it took an image of its foot pad so the engineers could see how it was situated.

"It's how I knew I wanted to work in space," he says. Seeing that image as a teenager blew his mind. They were on Mars. And now here he was, living his dream. With a blink of Phoenix's SSI, he took the same image. Pat says I should have a look.

"They're up on Sol Runner," he says. That's one of the specialized mission-planning software packages the team uses.

"I don't have an account," I say.

"Oh, you don't have an account?" He's surprised. How can I cover the mission without access to Sol Runner? Shrug.

RIGEL WOIDA, PAT'S SON, POPS INTO BOB DENISE'S OFFICE.

"I feel like we're in that sci-fi film. You know the one where the lone robot takes care of the habitat after all the humans have killed each other off," Rigel says. I think he's talking about *Logan's Run*.

Both Pat and Rigel Woida work as engineers on the Phoenix mission. "The apple does not fall far from the tree" is a saying people use in this situation. They're both big men with big personalities. Rigel is a bit more of a rabble-rouser than his pop. In one infamous Phoenix incident, Rigel nearly prevented the delivery of a little-discussed addition to Phoenix, the Organic Free Blank (OFB). The OFB is an inert test sample that is rigorously prepared and tested to exclude any organic material. It's a blank, an organic free blank. They can test the background levels of organic material with the OFB. If TEGA measures organics in a sample, they'll use the OFB to check if it's a false positive.

The addition of the OFB was much debated. There wasn't really money for it. But in the end, a plurality of scientists agreed it was worthwhile. So there was a last-minute scramble at the LPL to build one. It literally came down to the wire. Chris Shinohara and Rigel, along with the rest of the team, had been up all night putting its final paperwork together and preparing it for transport to JPL on the last possible day it could be delivered.

It took a little longer than they anticipated, and there was a mad dash to make their plane. They made a quick stop at their respective homes to grab some clothes for the trip. In his tired state, Rigel accidentally packed his toothbrush and underoos in a bag he'd taken to the gun range a few days before. Security at the airport was not happy with the shell casings in his bag. Somehow Chris managed to avoid Rigel getting thrown in jail and delivered the OFB to JPL in the nick of time. Another Phoenix crisis averted.

"WE JUST DID MIDPOINT, APAM, AND FINAL SEQUENCE RUN-THROUGH. It goes a lot quicker when it's just us," Bob Denise says. He needs a moment to focus. He wants to take one last look at the plan. Then Bob does the Command Approval Meeting (CAM). And that's it. He signs the official paperwork and the plan is pushed out into space.

I sit in the empty downlink room in the chair that's usually parked outside of the SSI office, "Pat's Porch." And, as you do on the porch, you contemplate. Or if it's late and you consumed heavily at a 4th of July party, you rant.

NASA, IN ITS 50-YEAR HISTORY, FORGOT WHY THE AVERAGE ANDREW B. Dreamer might love space. Going to space is difficult and bold. It's fraught with risk and insurmountable odds. We're not meant to be out there. Every voyage pushes the limits of our humanity. There is no routine mission to space. NASA failed the moment they let us think that. It all went flat when NASA stopped telling us that space is risky, nearly impossible. After we got over the excitement of the first moonwalk, NASA didn't move us to the next phase. They never upped the ante to give us more insight, take us farther down the rabbit hole. Either they got lazy or didn't think we could handle it. Either way, we were in the palms of their hands. They could have turned us into a nation of nerds, glorious math and engineer-loving eggheads. Instead they rested on their laurels, played it safe. They tried to keep telling the same story. First it was cute and funny, but soon we just wished they'd go away.

They lost us. If only they would listen to Dara Sabahi and Peter Smith. Those are the guys you want running the show—not this

Michael Griffin character. If Sabahi and Smith ran NASA like a tag-team wrestling crew, they wouldn't always be on the defensive about success and money.

One thing you start to notice in news coverage when you take an interest in the space program is how much each program costs. For Phoenix, it's $420 million. The figure is repeated over and over. Contrast that with how rarely you see the cost of the war in Iraq or Afghanistan printed in the newspaper. The Joint Chiefs are really good at selling war; if only NASA could sell space so well. Heck, we could have relocated the entire Baath party to colonize Mars. War and space solved. Bamn! If the story could be reframed in terms of the bold and geektastic, NASA might have a better than a snowball's chance in hell as not being seen as such futards. So here's what we want: Give us heroes who explore the universe, people engaged in an inherently risky business, and tell us about them. Show us their flaws, their drive, make us feel empathy and let us root for them. We want pioneers and risk-taking discoverers. We want to imagine the folks who are braver and work harder by pushing the limits of humanity. They exist. I see them working here every day. Don't hide them. Don't make it a once-in-a-lifetime opportunity to get an up-close view. Without people and a story to care about, we're left to shake our fists at the stupidity of a bureaucratic conversion error that causes a crash or a radar turned off too soon.

Instead of heroes, NASA gives us science goals and complex stories about the search for habitability. Those goals aren't bad, they're just uninteresting. How do I know? I dare you to tell me what was the science goal of the most famous of all the missions, the moon landing. You don't know because it didn't matter.

We need a new story with passionate risk-embracing characters to love. Outside of Buzz Aldrin and Neil Armstrong, most Americans would be hard pressed to name a third astronaut—even the one who traveled with Buzz and Neil on Apollo 11. Poor Michael Collins. In those exciting early days of NASA, they used to laugh in the face of risk. We want to laugh again.

CHAPTER THIRTEEN

DOWN AND OUT IN THE SOC

SOL 41

WE FAIL TO HEED EDNA FIEDLER'S ADVICE. AFTER A MONTH living on Mars time, the allure of a day off in the sun is too great. Instead of toughing it out, too many of us make a one-day shift to Earth time. If you've ever wondered what it feels like to go to Japan for the day and then fly back, I can now tell you: it hurts.

The team got just enough time off to remind their bodies how wrong this schedule is. Switching magnified the brain fuzz, aches, pains, fatigue, and general malaise—I could go on. But if you think today is bad, you should have seen us yesterday. We were in rough shape. But enough complaining, there's ice to scrape.

There's a new kind of badge holder here in the SOC. It's denoted HQ and it's worn by the enforcer NASA sent to watch our every move and report back. His name is Ramon de Paula. Somewhere in the building lurks the NASA program executive for Phoenix. He will ensure that we do NASA's bidding . . . or else. We will now work under the shadow of a NASA minder.

"They didn't approve the plan. HQ wants an official review," Peter explained to Leslie Tamppari yesterday when they arrived back at the SOC. He shook his head in disbelief—even though he was doing the briefing. The overlords back in Washington aren't ready to approve the scheme the team cooked up in the four-hour special Smith session. It's all postponed by official dictum until at least sol 44. Before sol 44 we are expected to plan and test for a new approach. Then once NASA feels comfortable, they'll give the go-ahead to open the TEGA doors.

Headquarters wants to be sure the Phoenix plan is really iceworthy. They want some guarantees—they love guarantees—the mission will get ice. Yestersol was a marathon planning session. The goal was a framework to convince NASA that the science team has a fantastic approach that will yield lots of ice. It went medium well. So we threw it out. Now there's a new plan with some newly devised tests.

"We're going to scrape about 80 times in the bottom of the trench," Ashitey says. Still, no one is sure about the effectiveness of the scraping. It doesn't look promising, and without better evidence our current plan is vulnerable to summary disposal. Right now it looks like the scoop just shifts around the dirt that's on top of the hard stuff. The sides of the trench fall in and dirt gets moved back and forth; this gives the illusion of scraping. That's not to say it won't work. They've only had limited time in this new trench. So they'll have to test and retest. Yestersol's plan asked Phoenix to deliver the dirt that was already in the scoop into MECA's optical microscope and wet chemistry lab. Carol was pleased it wasn't wasted. We should know soon if it worked.

Kickoff begins at just before 6:00 p.m. local Tucson time, around 2:32 p.m. on Mars. Curiously, there are lots of teenagers sitting on black plastic buckets in the back of downlink. That's not normal.

"These are students from a program called PSIP, Phoenix Science Internship Program. They are high-schoolers mentored by some members of the science team. They'll be here for a few days with their black buckets," says Cassie Bowman from the education and outreach department. The buckets have nothing to do with their education. There's just a chronic chair shortage in downlink; the buckets are supposed to keep the PSIP kids out of the way while they complete their summer internships and watch the wonders of space exploration.

"They will be shadowing their mentors and helping out. Please introduce yourselves," Cassie says.

"Welcome, students," Vicky Hipkin says. She is the sol's sci-lead. "So, today is a very big day. We expect the WCL success metric to be met if we get a sample in the drawer. We'll have a big cheer for that. And it's a big day in our progress for getting a TEGA sample," Vicky says. Usually I like Vicky's unflagging optimism and grand vision. But today it just feels like hype. She invites Aaron Zent, the strategic science lead, to talk through the strategic plan formed at yestersol's marathon session.

"In light of the fact that we've re-prioritized to get an icy sample, our strategic plan has changed. We intended to sample Snow White [at the Wonderland site] from top to bottom layers," he says.

Vicky pulls up the mission scorecard. It's the chart with all the boxes that tells you how your mission is faring in the grand scheme of galactic discovery. This is a favorite tool to boost spirits whenever things look grim. Doug used it just a few sols ago to keep us from walking out when NASA first came to town. The thing is, it's not a very pretty chart. They should have made one of those giant thermometers like the telethons have. Boxes with check marks just don't do it for me. Even so, the boxes indicate that things aren't going too bad. In spite of my foul mood and the long road ahead, the chart makes it clear that we make progress every sol.

In fact, the baseline success requirement for MECA is nearly met. They only have to get two samples for MECA and TEGA to have "minimum success." If the delivery works today, Phoenix will meet the MECA goal. Then they just need a sample for TEGA, and balloons will fall from the ceiling and Dick Clark will bring us champagne. The giant Peter Pan mosaic is almost complete too. The final images should come down in the first com pass. This is all good news, but who is going to be happy with "minimum success"? Not the kids on the buckets, and certainly not the overachieving scientists in this room.

"It's going to be a long day for Phoenix," Cameron Dickinson, today's SPI I, tells the team. There are four communication passes to handle all the data.

"That's the most yet," Cameron says. That's the spirit. We're going to stay up later and work harder. Minimum success is nothing. Now that the engineers feel more comfortable operating Phoenix and a larger library of safe activities exists, they're going to put more activities in the plan and Phoenix will stay up later and later. Even though NASA hobbled the TEGA effort, the other instruments should make up for what's lost. That's good for science, but puts more pressure on the team to code an increasing number of activities on the same short timeline.

"It's is going to be very difficult to work this plan," Vicky adds. But she's sure we've got the moxie to do it. Before Vicky ends the meeting, she asks for the weather report.

"We're storm-free. Everything is good!" Carlos Lange says.

Wait. Mark Lemmon says he has a question.

"Why are 72 images missing from yestersol's plan?" he asks.

An engineer from the spacecraft team responds hesitantly. He says there was an error with an APID. Something got mislabeled (read: someone effed up). Lemmon shakes his head. The images to complete the Peter Pan mosaic are lost. Now he must wait until there's available time in the sol plan to retake them to complete the image. We break.

John Hoffman, the TEGA co-investigator, asks the bucket brigade to follow him into the conference room. He's going to give a little lecture on TEGA to his crew of interns. I ask if I can join. He says sure. Hoffman's lab at University of Texas at Dallas did a lot of the fabricating for TEGA's EGA (Evolved Gas Analyzer). This kind gentleman is a controversial figure. There was a lot of strife and indigestion over the design and fabrication of an important part of the EGA, the gas ionizer. This piece of TEGA, sometimes called the ion-pump, helps move a sample through the EGA. The ionizer excites and speeds a neutral gas to carry the atomic bits through the instrument for measurement.

Bill Boynton shakes his head when I ask him about it.

"We thought that, together with JPL, we could hold their hands through the process," he says. "Heather asked NASA to assign a quality assurance specialist from NASA." Apparently they were out of practice and needed some help. Even with NASA stepping in, it didn't work out well. Much of the EGA had to be rebuilt at the last minute. Chuck Fellows from the Tucson TEGA group had to clean all these tiny parts by hand to try to get it working. It wasn't even clear if they would make the delivery date (or even launch). How it all went wrong is complicated. Most important is to remember that it's not easy to build an electron stripper (ionizer) for Mars. All that matters now is if TEGA can hold it together and keep delivering amazing data.

Hoffman's been building spectrometers since the seventies. And it doesn't seem like he's slowing down any time soon. He still runs an active Mars lab, mentors students, and is often here debating science when I'm walking out the door. It makes me wonder what kind of meds he got—something better than Provigil™, I suspect. (This is the only scenario that doesn't force me to confront the sad truth that this man has more staying power than I do.) We all sit down. Hoffman smiles sweetly at his students.

"There are six instruments on Phoenix." He starts his lecture with an easy one.

I knew it!

"Bill Boynton used a neutron emitter to show that neutrons interacted with hydrogen on the polar cap of Mars. And his analysis shows that these interactions took place about 5 cm below the surface."

Knew it!

"The two parts of TEGA are the 'TA'—the thermal analyzer—and 'EGA'—evolved gas analyzer." Knew it!

"Forty-five-degree slope on TEGA designed for extra dirt to fall off. And the small grate size is to keep the boulders out." Knew it! Knew it!

The lecture is fantastic. I take a moment to feel superior to some overachieving teenagers.

Hoffman doesn't go into any of the issues that cropped up with TEGA or the problems that plagued the ionizer. The students don't really grill him either. Now probably isn't the time.

IN THE RA OFFICE, ASHITEY MINDLESSLY CLICKS THE "REFRESH" BUTTON on his computer screen.

"What's taking so long to upload the images?" he asks no one in particular. Ashitey wants to know if they successfully delivered the sample left in the scoop to the wet chemistry lab, WCL. He's clicking refresh in the hopes that somehow he can will the images he is waiting for into the SOC. Not that they'd admit it, but I think they were personally offended that the sample they collected proved the source of this controversy.

"We worked really hard to get that sample," Joseph Carsten says.

"They wanted us to pick up 20 cc's!" Ashitey says in his most vexed voice. Although even at his most vexed, he has a calm, enlightened demeanor.

"That's like a couple of soup spoons. We've never even tested that in the PIT," Joseph says. "I am amazed that we did it. I couldn't believe we could pick that up. We got 15 cc. I didn't think we would get it. It was a great sample. Then, of course, the *President* called." That's Joseph's interpretation of how NASA intervened.

"We stayed all night working on that! Then they just held it. I'm glad they used it for something," Joseph continues.

"Did you know, we only know when current goes through the arm? That's like if you knew you were touching something with your finger based on the pain you felt in your arm muscles from pushing," Ashitey says. "We can just measure how much more current we're using. They wanted us to just go 1 cm into the ground. You can't make sensitive movements when your feedback loop only gives you a sense of how hard you're pushing!" He then demonstrates this by pushing his arm through the air.

"Look at how noisy the DEMs are," Ashitey protests. These are digital elevation maps that show the contours in the ground. "You can predict reliably in the air where you'll be, but once you touch the surface it's hard to know exactly what's happening. You don't have many senses to work with." Since the arm is flexible and swings, there are a lot of forces acting on it. That makes it hard to predict exact position.

"We warned the science team that it would be hard. We told them we only had a 50-50 chance. They don't really understand how difficult it is," Joseph says. Ashitey reminds me again that it's really important not to muck up the whole book by getting the RA wrong.

"I thought they were going to throw away that sample after the President called," Joseph says. They nod knowingly.

After hitting the refresh button a few hundred times, it works. Finally there are images of the delivery.

"It's got stuff in it," Ashitey says. He does a little on-the-fly color correction to the images to improve the view.

"Hmm. Now I'm not so sure," Ashitey second-guesses himself. He looks intently at the screen. From somewhere down the hall, clapping and cheering break his concentration. It's coming from the MECA office.

Ashitey and Joseph head over to investigate. They look happy and we crowd in their office. The cheering and smiling faces don't convince the RA team. They want proof.

"We should see if material spilled around the WCL drawer," Joseph says to a MECA engineer. They head back to their office.

"Well, if Mike Hecht looks *this* happy, there's nothing to worry about," Ashitey says. He changes his mind again and decides that yes, there is probably a working sample.

THE SOC IS FULL. THERE ARE 20 STUDENT INTERNS VISITING, A documentary crew, and some NASA HQ folks. I wonder if one of them is Ramon de Paula, our NASA minder. Ashitey and I stand in the back and wait for midpoint to start.

"Why are there so many kids sitting on buckets?" Ashitey asks. I tell him they're students being mentored by Peter and John Hoffman.

"Hmm, that's interesting," he says facetiously. "If I were you, I would consider putting that in chapter 10."

No, sir. Lucky chapter 13. But thanks for the advice.

INSPIRED BY DARA SABAHI'S PASSIONATE ADVICE ON RISK, I SEND Peter an email asking for specific permission to get into some senior management closed-door meetings. This short-circuit crisis makes me realize that life in the SOC is fleeting. It's time to act. I need to get inside to get the story. He writes back:

> The leadership caucus meetings are not open meetings.
> Peter

So much for that. I guess Peter thinks I'll get a better story if I have to work for my insider NASA secrets. Either that, or I'm doing it wrong.

THE SOC SECURITY DOOR FLIES OPEN. A MAN COMES RUNNING IN. "Where . . . is . . . Peter . . . .Smith . . . ?" He asks at full volume. There are a few uncomfortable laughs and then silence. "I NEED to find . . . Peter Smith." He breathes heavily even though he's only run about ten or fifteen feet.

Everyone looks around at each other. He pokes his head in offices, over cubicles, and around corners. No one is really sure what to make of this guy. "Ahhh . . . I think he's in a meeting," Mike Mellon finally says after a long awkward pause.

"CUT!" This guy says. He looks disappointed.

Sara Hammond escorts a TV crew through the SOC for some science show. "This guy, Josh, is supposed to be the next Crocodile Hunter; he's some kind of survival expert," she says—even though his Wikipedia page says he grew up in Manhattan.

"Please be mindful and stay out of people's way," Sara says to the crew. They're not paying attention.

"Geez, they don't really let things happen, do they? Sooo stagey," she says to me quietly.

The film crew tromps around. They continue to make lots of noise and get in the science team's way. Sara does her best to rein them in.

CHAPTER FOURTEEN

IN A SCRAPE

SOL 42

We are on a scraping mission to collect permafrost. Today's core activity paves the way for a TEGA delivery. A group gathers to look at the first scrape images. I push in behind them. The series of pictures makes a time-lapse film. The sequence shows material being pushed into a line. But then it disappears.

"Is it ice scraped up, or just dirt being pushed around by the scoop?" someone asks. One real possibility is that the ice is just too hard to scrape—even for a titanium blade. The images are inconclusive. Last night we stayed late debating how much documentation we would need to convince NASA that scraping in the bottom of the trench works. These are the images we're looking at now.

In the robot arm camera office, Morten Madsen, Walter Goetz, and Line Drube stare at the same scrape images. They will make a recommendation to proceed if they believe there is enough material in the scoop for a delivery. The SSI camera team does the same thing. Chris Shinohara comes into the RAC office.

"What do you think?" he asks.

"We're not sure yet. We need some time," Morten says.

Chris leaves. Morten, Walter, and Line stare. They speak to each other in Danish. Walter used to live in Copenhagen. So they're used to communicating that way. It's hard to know if they're happy or sad about what they see. But the smile-to-frown ratio isn't good. Chris comes back.

"We're all waiting. Post something!" he says.

"A priori, it's not looking good," Walter says to Chris in his heavy German accent. "Even if we have the material, I say, it's not enough for a delivery. Maybe we are wrong. That would be good."

"Well, I think there's sample in the scoop. And it's enough for TEGA," Chris says in disagreement.

"Unfortunately, our RAC divot images provide no help," Walter complains to Chris.

"Well, RAC and RA changed the exercise. It was not communicated to management, and the mission managers were not comfortable with the changes," Chris says sharply.

"That's a waste of resources," Walter says.

The fight is over a series of close-up images dreadfully out of focus. Walter and Bob Bonitz worked all day—when they should have been sleeping—to write the code for the images. They made a heroic effort to capture the closest images that the RAC can take. The plan put the scoop blade as close to the imager on the robot arm as the geometry and flight rules allow. Joel Krajewski decided the blade was too close to the scoop. If somehow a rock or something was sticking out from the tip of the scoop, it might damage the lens on the RAC. He deemed it an unacceptable risk. Just before the plan was finalized, Joel asked the sequencing engineer to tilt the scoop back and provide a margin of error. Moving the scoop changed the focal length.

"The images we spent our night perfecting are out of focus—worthless!" Walter declares.

This is where fatigue affects the mission—a breakdown in communication leads a scientist and an engineer to make different assumptions. It sets the team back a day. They would have had an easy decision if they'd communicated on these photos, but they didn't; so, regardless, we don't have them.

From what we do have, it's difficult to tell what's in the scoop. Are we seeing a shadow, a small new sample, or old dirt in the scoop? I'm glad it's not my decision.

"I'm sorry," Morten says. "Let's speak English so everyone can participate." I tell them I don't want to disturb.

"You can disturb us any time you need to," Walter says. They switch to English. Walter believes it's important for outsiders to bear witness to the mission. He's not afraid to say it, either. One sol he stopped me in the hall and put his hand on my shoulder. Then Walter looked deep into my eyes and said, "You are the most important person here. You make us . . . vat's the verd? . . . eternal . . ." And then he walked away.

Of course, when it becomes clear that there is no good choice here, frustration mounts. They switch back to Danish, where it's easier to colorfully express frustration.

Shinohara comes back a third time. He asks for a comparison between the images taken on sol 38 and today to see if there is new material or it's just old stuff that's stuck.

Ray joins. He and Chris decide they can't continue until someone makes an executive decision.

"Scraping works or it doesn't. I want evidence or an argument for or against," Chris says. The RAC starts to fill with mission managers and engineers.

"Technically, the images are great. But they don't help us," Walter says. Chris gives him a what-the-hell-does-that-have-to-do-with-anything look. Walter has a tendency to go off on tangents. He thinks there's a beauty in the mechanics of the image. Who wouldn't want to share that with colleagues? He can't help that he's a curious fellow. It makes him a good scientist and a fascinating guy. But there's a lot of pressure to get a plan finished and not a lot of time for musing. Chris loses patience with him. I'm not sure Walter even notices.

"Oh, maybe those might be particles," Walter says. Then he backs down. "Probably not." Either way, he still thinks the images are really great in spite of being useless.

"After 43 days living like this, you expect nerves to fray," Joel Krajewski tells me. We stand around waiting for someone to decide: go or no go. There's a lot of foot traffic between the RAC office and downlink. The problem is escalated up the chain of command. There's too much at stake and not enough evidence. Now it's up to the mission managers to decide if the test was a success and ask NASA to move ahead.

Nilton Renno asks how many scrapes they made. Joseph Carsten says there was a total of eighty.

"Maybe we made a discovery of some really hard new material," Nilton Renno says. Joseph and Ashitey give Nilton a blank stare.

"That was a joke," he says dryly. He laughs.

"Oh, I have an idea," Nilton says to himself and walks off.

"This is going to be exciting," Joseph says.

NILTON RENNO'S BIG IDEA IS NOT A BREAKTHROUGH DISCOVERY. IT'S a new joke. He wants to include it in a presentation he's working on. But before he springs a joke on a larger audience, he wants to test its humor value.

"No one is really laughing yet, but I think it could be funny," he says. The rest of the SOC might be in crisis mode, but Nilton has something else in mind. He's like Columbo: you think he's on the completely wrong track and then he surprises everyone.

"I think what's missing is a good image of an obelisk," Nilton says. "That will sell the joke," he says. After a quick search, he finds his image. Nilton heads to the printer to collect his printout. Leslie Tamppari is standing at the printer. She volunteered to be subject number 1. Nilton gives it a go. He shows her the obelisk. She doesn't laugh.

Nilton heads back to his computer. Tenacity, if not humor, is one thing he has in spades. He searches for a better image. Finally, he's happy and heads back to the printer.

"I found the problem," he tells me. "Now I reversed the key events in the build-up to the punch line." Now he needs a new subject. He walks over to Mike Hecht; he shows Mike piece of paper number one. Then he shows him the picture of the obelisk from 2001 at the Phoenix dig site. Get it? It's like we've just found the buried obelisk and that's why it's hard and black and we can't scrape it up.

"Haha!" Hecht laughs out loud. Success! Nilton smiles and returns to his desk.

KAREN MCBRIDE TELLS ME I NEED TO TALK TO RAMON DE PAULA, THE mysterious NASA heavy lurking somewhere in Mission Control. I scan badges every day to look for Ramon, but nothing yet. Karen worked with Ramon de Paula and Bobby Fogel to oversee the mission for NASA. But she quit after a difference of opinion created a rift between them.

"They don't care about science. It comes down to political appointments and ass-kissing. It's a miracle Phoenix got done at all. It's by the grace of the team's hard work," she says, railing against her former colleagues.

"Bobby Fogel and Ramon de Paula never even once called Peter for a sit-down before going to the top about the TEGA concerns," Karen McBride says. She's not the only one to point out this omission.

She objected to how they were doing business and refused to work with them. Now she's here for moral support. That makes her Ramon's foil, fighting to keep NASA on the up-and-up.

"Ramon is here too," Karen says. She doesn't want me to just hear one side.

"Get the whole story, talk to Ramon," she says. "Be careful, though."

PETER AND JOEL DISCUSS THE PROCESS FOR MOVING FORWARD. HQ is going to want a new plan. Should they just rasp? Rasp and scrape? These all need to be tested. They'll spend the rest of the day outlining a plan, and then Peter will make sure that NASA is on board.

Chris is busy trying to figure out why the RAC went into safe mode after it took the images they were arguing over earlier.

"There may have been an error in the sequence," he says. He starts to explain what went wrong when Heather Enos interrupts.

"This is not funny, Chris," she says. She holds up a sheet of paper. It reads "TEGA Memorial." At the top, the word "Report" had been replaced with "Memorial."

"It's kind of funny," Chris says. Heather walks away. Chris says midpoint will happen on schedule. Shinohara goes to see if the RAC and SSI teams are ready.

Midpoint starts at 10:05 p.m. local Tucson time. It's 4:01 p.m. on Mars. Nearly on time. The SSI team reports that the images don't show much of a pile in the scrape documentation. Walter Goetz gives the RAC opinion. He agrees that there's not enough material for a delivery.

"Engineering-wise, we took a great image," he says. "We planned four images, but we have no way of knowing these particles weren't there before. The instrument is healthy." Mission managers agree, the material is too hard to scrape. They will not get enough material for TEGA if they only scrape. It's time to use the rasp. Over the next week, they'll prepare the little drill at the end of the scoop to excavate permafrost on Mars. Bob Bonitz thinks it'll take about seven sols before they're ready. For now, they'll complete the wet chemistry experiments and prepare the atomic force microscope.

NILTON RENNO GIVES HIS PRESENTATION. IT'S ABOUT DUST STORMS on Mars. We learn a bit about how surface material moves around the planet. Then the big moment comes, the joke. Set up. And cue the obelisk. . . .

He gets some laughs. Moderate success. It's Nilton's effort to lighten the mood in the SOC.

"I heard people under stress don't like jokes," he tells me later. "So it was a little experiment."

The joke may only evoke a moderate response, but the dust talk sparks a conversation about the center versus the boundaries of polygons. This is a recurring debate about what makes a good sample. Which begs the question, how do we discover Mars. It's a calm interaction. Everyone agrees that no matter where we look, one thing is certain: polygons on Mars are great. No debate. No arguments. They are great.

CHAPTER FIFTEEN

POWERS OF TEN

SOL 43

THE SOC IS MOSTLY EMPTY TODAY BEFORE SHIFT. NILTON CHATS in Portuguese with a nervous-looking fellow who looks kind of familiar. Is this the Brazilian love-child of Peter Lorre and Steve Buscemi . . . here . . . talking with our Nilton Renno?

"Have you met Ramon de Paula?" Nilton asks me.

No. This guy? He is the enforcer NASA sent to do its dirty work. Ramon is not nearly as intimidating as I imagined. His fidgeting and small frame don't match the image of a tough guy I'd constructed. Nilton tells Ramon I'm doing a "special project" for Peter to document the mission. Does he suspect that I'm a rogue infiltrator here to sneak a peek at the soft, fleshy underbelly of mission life? Nah.

"It's important to get the story out. I'm glad you're here to help," he says. He's so friendly. Maybe it's a ruse to draw me out. I've seen *Three Days of the Condor*, you sly devil. I know how this works.

"It's important to understand Mars in order to understand Earth. They were once very similar and now they are so different," Ramon says. He says there are a few books on the subject that really inspired him. I'm so desperate to look interested and figure out his angle, I can't remember a single title he mentioned.

Ramon de Paula is the NASA program executive whom Karen McBride told me to talk to, and here he is. How can I be enemies with someone who is so nice? Nilton and Ramon chat for a bit, and then he walks away.

Ramon is here to get the dirt on TEGA. Cordiality be damned. Maybe he is happy I'm here, and Peter is doing the right thing with his media outreach. Nevertheless, I'm going to suspect the worst so as not to complicate my cosmology of NASA. I can't rewrite that whole July 4th rant. He must be here early trying to chat up Nilton and squeeze some info out of him. Unfortunately for him, Nilton doesn't engage in gossip. He just tells you what he thinks. And that's what gets him into trouble.

"WE HAVE A GOOD DAY AHEAD OF US. I'D LIKE MARK LEMMON TO SHOW something," says Doug Ming, today's science lead.

"This is the full mission success and deck pan," Lemmon says. It's the entire 360-degree view of Mars stitched together from hundreds of high-resolution images. Looking out over the lander deck is the sparse rock-speckled landscape stitched together in full color. This is my chance to lose myself in the Martian ideal. Uneven bits of rocks, with sharp edges, poke out of the polygons. It goes on forever. The reddish-brown rock-speckled landscape looks like toasted almond pieces poking out of a milk chocolate shell. The Martian arctic is a Dove bar. Everyone claps.

"Good morning. I'll be your shift lead. Now let's hear from our TDL," Richard Kornfeld says, eager to get started. Our data manager is Jim Chase. He speeds through a list of data packets and error reports. There are a few SSI images missing, but everything seems to be healthy. Jim recovered nicely from his tangle with Julia Bell and doesn't miss a beat. Data should arrive in fifteen minutes or so. We get instrument updates and break.

THE MECA OFFICE CHEERS ABOUT SOMETHING. IT'S THE TECP. WELL, A little movie of the TECP needles inserted into the soil. Sticking just the needles in the dirt without jamming them is not easy. Everyone seems a little surprised that it worked. We congratulate Ashitey with kind words and handshakes. We need something to celebrate.

"Management didn't have a lot of confidence you could do it. Even with noisy DEMs from SSI, you got it," Deborah Bass, the Deputy Project Scientist from JPL, tells Ashitey.

At midpoint, there's a new surprise.

"We got 15 more megabytes of data on board than predicted. So some things will be lost; mostly SSI images and some MECA data will be lost in the buffer," Jim Chase says during his data report. Considering Phoenix only has 100 megabytes for storage, that's a lot of extra data. And no one really knows where it came from. More data seems like a great thing. It's not. Great care goes into tracking these bits and bytes; extras means something is amiss. Plus, this overgrowth pushed out some of the images scheduled for downlink. They're gone forever.

"What kind of disconnect is there that gave 15 MB too much data?" Richard Kornfeld asks.

The spacecraft team might want to restrict the science plan until they know why they are generating more data than predicted. Apart from Phoenix seemingly doing activities on its own, everything else is normal. I suspect the ghost in the machine objects to the changes NASA implemented and is going ahead with its original plan.

The RA team will continue work on scraping techniques. MECA will continue to process their data. And we'll keep waiting until we get ice inside TEGA.

TWENTY MINUTES BEFORE THE END-OF-SOL SCIENCE MEETING, PETER meets with his scientist-in-training interns. They're all laughing and talking about living on Mars time.

"It's not so bad," one of the students says.

"Good, I'm glad you guys are enjoying yourselves," Peter says gently. "Let's talk about your project."

"These students are here to do *real* work," Peter tells me. He didn't want them to come all the way to Mars for make-work. So he's constructed actual projects. One team makes Mars dirt samples based on the new chemistry findings. This group will make a film. It's a Mars version of *Powers of Ten*, a classic movie about the beauty of scale by the designers, Charles and Ray Eames. Peter wants them to examine the powers of ten on Mars.

One of the highlights of Phoenix is that it can image huge panoramic vistas, atomic-force nano-pictures, and everything in between. Making interesting "data products," as the scientists call them, is important work. Usually the camera team makes these kind of layman-friendly images that are a big part of how Peter wants the public to connect with the mission. On Phoenix, it's been hard to put out anything more than raw images.

In spite of the time crunch, mosaics and comparative images end up on the Web. Amazingly, it's the public that picks up the slack. There's a group of dedicated fans and skilled amateurs who post their work to fan sites. They do it just for the love of Mars.

"We should be doing as much as we can to encourage that," Mark Lemmon told me one afternoon over coffee. "These are the people who get excited and write their Congressmen to keep us in business."

"Don't limit yourselves to Phoenix either," Peter says to his students. "Look at HiRISE to get even larger-scale images." They're excited about the project—Peter and his students. They cook up a plan and then idly chat.

"Have you heard about the perchlorate?" Peter excitedly asks his students. They shake their heads. I shake my head too. I don't even know what perchlorate is. Sam Kounaves, the MECA Co-Investigator, sits nearby and overhears the discussion. He looks over, a bit shocked.

"We haven't announced that yet," Sam says. Is this the big secret the MECA team is hiding!? That would explain why he's annoyed.

"Well, what are you waiting for?" Peter asks.

"Tomorrow . . . ?" Sam says with some hesitation. He definitely doesn't want to be put on the spot.

"We still have data coming back . . . and we need to prove it," he says.

"Always tomorrow!" Peter says. "I'm going to give your talk if you don't."

Later that day, Peter gives an interview to a local PBS affiliate. They ask what is the most exciting thing he's seen on the mission.

"I can't tell you," he says with a fiendish chuckle.

I do a Google search for perchlorate. I discover it's a chlorine-based ground water contaminant. Contaminant!? It's highly toxic to humans, poison. And it can be used as rocket fuel. Rocket fuel? I should follow up with someone on this.

DICK MORRIS STANDS AT THE FRONT OF THE CONFERENCE ROOM.

"I'm going to start," he says quietly. Everyone talks over him. He is always so civil. The team settles down. He puts up some graphs of the multispectral images of the ice. He shows images from the sol 31 and 32 digging sequences.

Based on his reading of the photos and data from the RA, he thinks there is a lot of dust in the ice they're trying to scrape. He discusses the data and makes a long trail of logical steps with sound scientific principles. (I lack the tools necessary to follow these steps, but Morris has such a friendly, trustworthy look, I can only assume they're spot on.) It leads to a conclusion that this ice is not going to be easy to scrape.

"If the goal is to sample ice, we need to generate a large volume of dark-toned material and rapid-transfer it to TEGA," Morris says. He explains the ice will be too hard to scrape. And based on the data, rasping is the only hope for getting a sample. It's food for thought and justifies the decision to abandon scraping and work with the rasp.

"You've got too much time on your hands, Dick," Bill Boynton says jokingly. "You need a real job."

CHAPTER SIXTEEN
NILTON'S NODULES

SOL 47

I'S AN HOUR BEFORE KICKOFF, AND THE SOC IS MOSTLY QUIET.
Wandering around, looking a bit lost, is Ramon de Paula. You might
mistake it for the look of a man who has missed his bus and doesn't
know when the next one is coming. At the SOC, it can only mean one
thing: Ramon is not on the "phx_surf_ops" email list. We can only
guess how long he's been here.

The team kindly gave Ramon a desk and workspace, way in the
back of downlink, near the kitchen in the back of the SOC. Every
five minutes or so, he wanders out, does a lap around, and heads
back. Ramon can't understand why the SOC is empty and kickoff
isn't starting. He probably calculated the start time for today's sol
just like I did on my first week—adding forty minutes. Little did he
know that's just the simple math. Tired of the wandering, he swal-
lows his pride and makes his way over to Bill Boynton's desk. Bill is
here early, analyzing data.

"How do I find out when kickoff starts?" Ramon asks. He's having
the same issues adjusting to life in the SOC as I did back in the day.
How cute. Ramon and I are two peas in a pod. Ramon doesn't like

that he's not on the "phx_surf_ops" list. *Talk to Sara Hammond, she'll get you sorted out.*

"Be careful," Nilton says with a raise of his eyebrows. What is that supposed to mean? Ramon will send me packing? How did he even know I was watching him? Am I that obvious? (This space book is taking a turn for the Skull and Bones.)

"Come take a look, I have a theory about the barnacle," Nilton says before I can press him on his Ramon warning. We look at more images of Snow Queen, the icy block under the lander. There's no time for faux conspiracy: there's science to do. This particular picture is a blow-up of the lander legs. Nilton superimposed some markers to highlight an area on the lander legs.

"These particles are growing, and I think I have an idea why," Nilton says. In the images, you can see several particles move around a bit and get larger. The time-lapse photos make it all look like a chaotic barnacle ballet.

"This might be a briny solution splashed onto the legs at landing," he says. "This could be a deliquescent salt. Maybe a carnallite," he says and writes down the chemical symbol $KMgCl_3 \cdot 6(H_2O)$. A deliq-what? I don't know what that is.

"I'm going to give a talk about it today," he says. He'll explain.

KICKOFF BEGINS.

"We'd like to build a 'super strategic plan' to figure out how we're going to sample until sol 90," Ray Arvidson says.

NASA officials decreed that Phoenix has to indicate their intentions for digging. They want to see a plan from sol 48 to 90, the rest of the primary mission.

"And we want to better inform NASA," Ray says with a smile. The super strategic plan will offer guidelines where and when the digging takes place for the rest of the primary mission. I guess HQ and Ramon feel left out.

"This is a very long day," says Suzanne Young, the SPI I. Phoenix is awake for 21 hours. The atmospherics theme group (ASTG) will make coordinated observations with the orbiters circling Mars late into the night. Christina von Holstein-Rathlou, one of the Danes in the atmospherics group, gives a weather report before we break.

"This might be the first day that it's nicer on Mars than Tucson," Christina says. "Clear. No storms." We break.

I overhear someone talk about a party at Peter's place. All the Co-Investigators and JPL folks are going over for cocktails after shift. No, I'm not invited. Yes, it stings a bit. Note to self: be more charming and slightly less intrusive.

THE ASTG TEAM MEETS WITH LESLIE TAMPPARI.

"We're planning for another coordinated science day on sol 55," Leslie says. They'll build a "straw man" plan and divide the activities among the team members. They'll use this hypothetical plan to secure future space in Phoenix's schedule. These coordinated observations take a long time to build and require cooperation with the science teams who operate the orbiters.

"I'd like to look at the nighttime science," Leslie says. There are dozens of activities they'd like to do after 10:00 p.m., but they're a little unclear on how to heat the instruments. The pre-heating requirements for the instruments are complicated. If you don't pre-heat your instruments properly, they could blow a fuse. Changing fuses in the basement is annoying. Changing fuses on Mars, *very* annoying. If you don't want to violate the flight rules or destroy your instrument, you need to build math models based on the weather data you've gathered thus far on the mission.

"We're working on the problem," Joel Krajewski says. They're making heating models that should simplify the flight rules for the science team.Calculating pre-heating requirements is confusing and unclear to the ASTG group. They're not alone. It's same problem many of the groups have as they start to operate their instruments at night and early in the morning. It's one of the major stumbling blocks for getting ice into TEGA, too.

"That's wonderful," Leslie says. The other co-investigators agree. The ASTG group is really excited about Joel's heating-parameters talk. Once they get it sorted out, they'll have 52 new observations.

John Moores interrupts the ASTG meeting.

"I just want to let you know that the RA safed," he says. John is the strategic SPI. It's his business to know what the groups are working on. The strategic SPI builds a plan for the day after the day that's currently being planned. Today we're planning for sol 48 and he's working on sol 49. He gives ASTG a heads-up because this delay will push out their next coordinated set of observations. Their activities get bumped. And they don't like it. Someone suggests that they take a stand and fight for their observations. It requires a lot of work to coordinate the timing of these observations with the team that operates the Mars Reconnaissance Orbiter. And they don't want to repeat the process.

IT'S NOT JUST THE RA THAT SAFED. THREE INSTRUMENTS—THE RA, TEGA, and RAC—have all gone into safe mode. Three at once has to be a new record.

A young engineer named Katie Dunn is the tactical downlink lead. This is Katie's first Mars mission. Now she has three crises to deal with tonight. Depending how you look at it, this is either a peptic ulcer-inducing moment or a great opportunity. On a big mission, it's unlikely that a young engineer like Katie would get a chance to deal with these kinds of problems. On a larger, well-funded mission, this crisis would be handed by a specialist with a bit more experience. But said specialist isn't here and the problem is left in Katie's capable hands. She works under Julia Bell, who is more than confident that Katie is up to the task.

"I'm not too frazzled yet," Katie says. "And, guess what? I saw the back of your head on the news last night." Finally, my mission chronicling gets some of the fame and notoriety it deserves!

Okay, no more small talk. Katie asks the RA team to join her while she scrutinizes a document called the predicted event file (PEF). She will use the document to determine which the commands the spacecraft received moments before the RA shut itself down. Then they can reconstruct the incident and figure out what happened.

To add to Katie's stressful evening, she's being shadowed by an even greener engineer, Joe Stehly. Katie has him talk through the problem as a sort of training exercise. She's not only taking care of business, but finds teachable moments in her stressful day. Yes, she does it all; Katie is the engineer of your dreams.

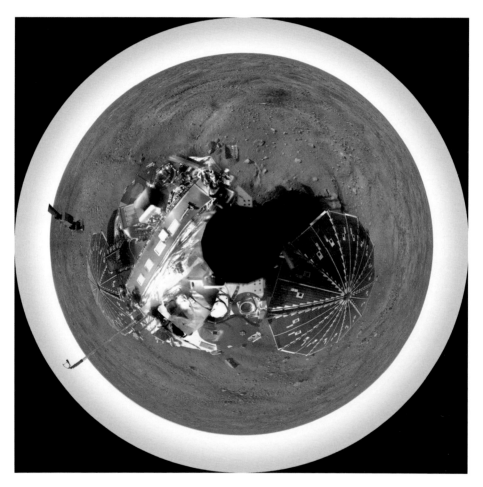

Not an official member of the Phoenix team, James Canvin is one of many Phoenix image enthusiasts who produces his own mosaics. This is his version of the "mission success pan." *Credit: James Canvin.*

The Author and Mads Ellehøj contemplate the SOC after shift. *Credit: Line Drube.*

ABOVE LEFT: Barry Goldstein (Phoenix Project Manager at JPL), Ed Sedivy (Spacecraft Manager, Lockheed Martin), Peter Smith (Principal Investigator), and Fuk Li (Director of Mars Exploration at JPL) celebrate a successful landing. *Credit: JPL.* ABOVE RIGHT: Dara Sabahi (Chief Engineer for Phoenix) and Charles Elachi (Director of the Jet Propulsion Lab) congratulate one another. *Credit: JPL.* BELOW LEFT: Richard P. Kornfeld (Senior Engineer responsible for landing communications) celebrates after touchdown. *Credit: JPL.* BELOW RIGHT: Phoenix engineers after landing. *Credit: JPL.*

Bill Boynton shakes his booty to celebrate the first successful acquisition of dirt on sol 16. Peter Smith and the team look on and applaud the momentous occasion. *Credit: John Beck/JPL.*

Peter Smith, Bill Boynton, Vicky Hipkin, and Mark Lemmon deliver the news that they discovered ice on Mars. *Credit: Line Drube.*

Peter Smith's captain chair sits empty as engineers and scientists plot the day's course. *Credit: Line Drube.*

The author sneaks into the Phoenix team photo. *Credit: University of Arizona.*

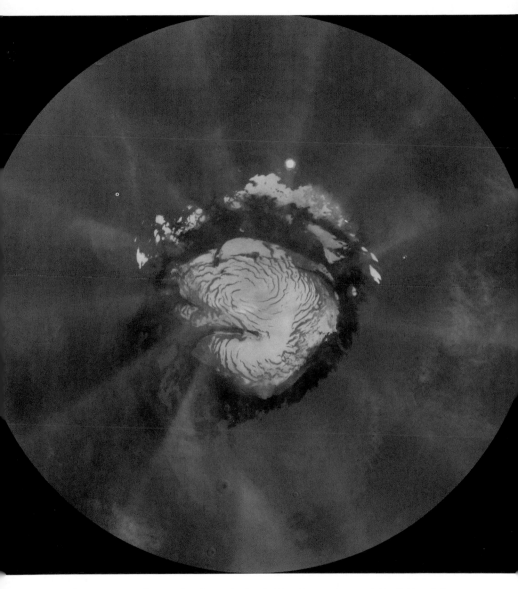

This is an image of Mars taken from orbit by the Mars Reconnaissance Orbiter's Mars Color Imager (MARCI). The Red Planet's polar ice cap is in the middle of the image. *Credit: NASA/ JPL-Caltech/Malin Space Science Systems.*

The Phoenix Mars Lander, with solar panels outstretched, gets a pre-launch inspection in a NASA clean room. *Credit: NASA/JPL/UA/Lockheed Martin.*

NASA's Phoenix Mars Lander is lowered into a thermal vacuum chamber at Lockheed Martin Space Systems, Denver, in December 2006. *Credit: NASA/JPL/UA/Lockheed Martin.*

ABOVE: The Robot Arm Camera maneuvers itself under the lander to capture the feature called Snow Queen in this color mosaic. *Credit: Kenneth Kremer, Marco Di Lorenzo, Phoenix Mission, NASA, JPL, UA, Max Planck Institute.* RIGHT: This mosaic assembled from a series of photos taken after Snow Queen piqued the science team's interest. Splotches of Martian material on the landing leg strut at left could be liquid saline-water. *Credit: Kenneth Kremer, Marco Di Lorenzo, NASA/JPL/UA, Max Planck Institute.*

This mosaic documents the midnight sun during several days of the mission.
Credit: NASA/JPL-Caltech/University of Arizona/Texas A&M University.

This image shows a soil sample from a trench informally called "Rosy Red" after being delivered to a gap between partially opened doors on the lander's Thermal and Evolved-Gas Analyzer, or TEGA. *Credit: NASA/JPL-Caltech/University of Arizona/Texas A&M University.*

TOP: This images shows a thin layer of water frost visible on the ground. The frost begins to disappear shortly after 6 a.m. as the sun rises on the Phoenix landing site. The rock in the foreground is informally named "Quadlings" and the rock near center is informally called "Winkies." *Credit: NASA/JPL-Caltech/University of Arizona/Texas A&M University.*
BOTTOM: The Surface Stereo Imager on NASA's Phoenix Mars Lander caught this dust devil in action west-southwest of the lander at 11:16 a.m. local Mars time on sol 104. Dust devils have not been detected in any Phoenix images from earlier in the mission. *Credit: NASA/JPL-Caltech/University of Arizona/Texas A&M University.*

ABOVE: The Phoenix editorial team selects images for press release. *Credit: John Beck/JPL.* BELOW LEFT: This is a view of NASA's Phoenix Mars Lander's Robotic Arm Camera (RAC) as seen by the lander's Surface Stereo Imager (SSI). *Credit: NASA/JPL-Caltech/University of Arizona/Texas A&M University.* BELOW RIGHT: This is an extreme close up of Martian regolith taken with the Optical Microscope (OM). This sample was taken from the top centimeter of the Martian soil. *Credit: NASA/JPL-Caltech/University of Arizona/Imperial College London.*

ABOVE: This 3D view from the Surface Stereo Imager on NASA's Phoenix Mars Lander shows the trench informally named "Snow White." This anaglyph was taken after a series of scrapings by the lander's Robotic Arm on the 58th Martian day, or sol, of the mission (July 23, 2008). *Credit: NASA/JPL-Caltech/University of Arizona/Texas A&M University.* BELOW: The sun comes up on another Martian sol. *Credit: NASA/JPL-Caltech/University of Arizona/Texas A&M University.*

ABOVE: Phoenix tests out the sprinkle method for delivering dirt to TEGA. The scoop is held at a steady angle while the rasp vibrates, causing the dirt to flow off the edge of the scoop head. *Credit: NASA/JPL-Caltech/University of Arizona/Texas A&M.* BELOW: This image was taken by NASA's Phoenix Mars Lander's Surface Stereo Imager on sol 11 (June 5, 2008). It shows the Robotic Arm scoop containing a soil sample poised over the partially open door of the TEGA's number four oven. *Credit: NASA/JPL-Caltech/University of Arizona/Texas A&M University.*

ABOVE: This image shows soil delivery to NASA's Phoenix Mars Lander's Microscopy, Electro-chemistry, and Conductivity Analyzer (MECA). *Credit: NASA/JPL-Caltech/University of Arizona/ Texas A&M University.* BELOW: This image shows Martian soil piled on top of the spacecraft's deck and some of its instruments. Visible in the upper-left portion of the image are several wet chemistry cells of the lander's Microscopy, Electrochemistry, and Conductivity Analyzer (MECA). *Credit: NASA/JPL-Caltech/University of Arizona/Max Planck Institute.*

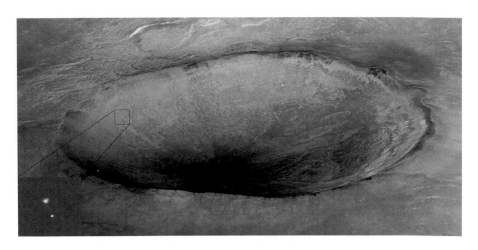

Taken from the HiRISE camera on MRO, this is a first-of-its-kind image: one spacecraft capturing the final descent of another spacecraft onto a planetary body. The insert shows Phoenix dangling from its parachute just two minutes and fifty-two seconds away from landing on the Red Planet. *Credit: NASA/JPL-Caltech/University of Arizona.*

Bob Bonitz and Richard Volpe test the rasp on the robot arm (RA) in the Phoenix testbed. *Credit: NASA/JPL/University of Arizona.*

The Surface Stereo Imager captures the Phoenix workspace and its own shadow on sol 90.
Credit: NASA/JPL/University of Arizona.

"There's no fault path," Joe says. A fault path indicates the RA recognized some trouble and shut itself down. They don't see one, so some sudden event caused the RA to safe. "There wasn't enough time elapsed for a timeout. So it must have been a sudden arm problem," says Rich Volpe, an RA engineer. That's a bit more alarming. They look carefully through the PEF log to find exactly what time the safe command was executed. It's almost midnight local time in Tucson and late afternoon on Mars. No one knows for sure what the problem is yet.

"It might be a power issue," Jim Chase says. He dodged a bullet by not pulling TDL duties tonight. Even so, he's keeping tabs and lending a hand when necessary. He doesn't think it'll be necessary, however.

Dave Hamara from TEGA pops his head into the TDL office.

"We think we didn't pre-heat the EGA enough," he says. He wants everyone to be clear this is just an initial thought.

"Twenty minutes would not be enough if the temperature dropped for some reason. We think. But we're still working on it," Dave says.

Ray Arvidson asks everyone to gather quickly for an update.

"RA, RAC, and TEGA are off the table. The SPI needs to quickly pull together a remote sensing day," Ray says. They'll try to salvage some science. A lost day is not what we need. Ramon de Paula comes in and asks Ray what safed. Watching him slink around and ask everyone what's going on really makes me feel the strength of our connection. I wonder if he feels it too. Maybe I should show him the ropes. He can use my patented Mission Control information-gathering strategy. We'll be just like Starsky and Hutch.

I head back over to the TDL office. Katie shows Joe, her TDL in training, how to search for different command strings in the blocks. This is my big moment to learn how to do a bit of TDL–ing myself. A search command here, a PEF query there, and soon they start pulling together the pieces. Katie has things up and running in no time.

Kristoffer Leer and Line Drube are the team members responsible for tracking down the problems with RAC.

"We're having trouble figuring out what's going on and what to do. We can't see anything," Kristoffer says to Katie when she gets a moment to check on them. The reason that they can't see anything is because they're ITAR-restricted. Katie groans, then pauses to think for a moment. Now she has to explain what's happening with the RA without violating federal law and risking a long prison sentence. No problem. Then she must act as the ITAR bridge to determine why the RAC safed. More work.

"The RA might have hit something. We're not sure what, but probably the ground," she says. There is no ITAR raid.

Morten Madsen, Kristoffer and Line's boss, comes in to help.

Katie gives him a brief update. And explains that the RA hit something. Could the RAC camera be damaged in an impact? The RAC is attached to the forearm of the RA.

We all imagine the RAC dragging on the Martian terrain. It's a horror of twisted metal and glass. It would be a sad end to our Mars imager.

"I don't think it's possible that the RAC can hit the ground," Morten says. Phew.

"So what shall we report at midpoint if we can't see anything?" Kristoffer Leer asks. She's not sure. She'll figure it out and get back to them.

Katie continues her rounds. She's got the RA and RAC in triage. Now TEGA needs some attention. She thinks Dave Hamara has it right about the temperature issue. Chris Shinohara comes into the TDL office. He is a welcome face.

"There is going to be an anomaly response team," he says. "You're on RA and RAC response. I'll find an engineer to sort out TEGA."

Thanks.

"We think the RA issue is pretty well understood. The scoop hit something," Chris says. Katie says she's already on it. Unfortunately, they don't know what or why. They don't want to unsafe it until they know where it is. They do want to unsafe RAC and get some images of the scoop to see what it might have hit.

Katie marches over to the RA office. She confidently strides across downlink and she's really hot—*Wait . . . What?* You can't say the engineers are hot in a science book! (Mars Lag symptom 93: Inappropriate Mission Control thoughts. *Focus on the mission, dammit.* I wonder if astronauts ever make out in space. For shame. This is a professional Mars mission. FOCUS.)

"Okay, what you boys got?" Katie asks the RA team.

"We hit a rock," Matt Robinson says.

"And . . ."

"The planned move has a trajectory and is plotted through various 'via' points," Matt explains. The RA team uses several methods to command the arm. In this version, they plotted a course for the arm and gave it some mile markers for the path it was supposed to follow.

"This worked fine in our simulation. But on the spacecraft there were probably a few 'encoder tics' difference and it hit a soft stop. Then it did an onboard recalculation and went through some wrist motion to avoid the soft stop. Unfortunately, it hit a rock on the way," Matt continues his interpretation of what happened.

A soft stop is a software limit that prevents the RA joints from moving too close to their maximum range of motion—hard stop. It can be dangerous to get too close to these hard stops, so the computer limits the RA from going to crazy when it gets excited. There are slight differences in how the robot arm moves in the test bed and on Mars. In this case a few millimeters difference could have been enough to cause the problem.

"As simple as a rounding error in an initial position," Matt Robinson says. This difference might have moved the RA too close to a soft stop. That error was likely tiny, maybe even one "encoder tic." (That is one increment on the gear that controls the joint.) The arm is just smart enough to know that if it's going to approach one of these positions, it should try to re-calculate its trajectory to avoid it. It does this before it even starts retracting its TECP needles from the ground. In this case, it found a better route. It rerouted itself.

Unfortunately, it might be smart enough to find a new path, but it can't see anything in its way. Ironically, the safety precaution led to it hitting a rock.

"How do we prevent this from happening again?" Katie asks.

"It's kind of a freak occurrence, so I wouldn't be too worried. But we could provide extra soft stop margins," Robinson responds.

"What we do need, before we can proceed, is high-resolution stereo images from SSI. The arm could be still torqued up against the rock. Carelessly turning the RA back on could damage a joint," Robinson explains. He wants to consult with an SSI engineer to make sure the RA scoop is in view of the camera and they can get the right image.

Okay, one problem is now well understood. What next? Katie moves back to the RAC issue. That story is coming into focus. The RAC safed because the RA safed. Joe Stehly found a RAC command that called on the RA. Since the RA was already in safe mode, the command went unheard. Without the proper response, the RAC did the only thing it knew how: it too went to safe mode. Easy.

Unfortunately for John Moores, who is the strategic SPI for the day, the plan he was working on for sol 49 is no longer valid. Delete. He has to start over. He'll re-do his rounds, polling the theme leads and instrument teams.

On my way to see the first images of the crash, I pass Nilton.

"Don't forget to attend the end-of-sol [science meeting]!" he says. It's going to be really exciting.

I won't.

Everyone squishes into the RA office to get a look at the accident scene. The skid marks from the rock crash tell the tale of our first collision. It doesn't look like there's any damage.

"It's our first rock-moving experiment/traffic accident," Mark Lemmon says.

"That wasn't a validated activity," Mike Mellon says. "There's probably going to be a penalty for that."

"It's validated now," Lemmon responds. Planetary scientists doing Vaudeville. There's going to be a quick midpoint meeting to assess the situation, and then a scramble to get a few new activities together and unsafe the RA.

IT'S 1:16 A.M. LOCAL TUCSON TIME AND 5:53 P.M. ON MARS.

"Welcome to a new and exciting day at the SOC," Ray calls the meeting to order. "Shift lead, please tell us what's going on."

Chris Shinohara explains the situation to the shift II personnel: three instruments are safed and we're rapidly updating the plan. He asks Matt Robinson to explain what happened to the RA.

"The action occurred when we retracted the TECP needles," Matt explains. "When the spacecraft executed the sequence, it hit a soft stop and calculated a complex set of wrist moves to avoid it. It led to the collision," he says.

"Is it possible we damaged the TECP needles?" Chris asks.

Matt doesn't think it's likely.

"TECP probably did not impact the ground, but there is a desire from the RA team to image the rasp to be sure it's healthy," Matt says.

Mike Hecht, the MECA co-investigator, isn't happy with that response. He'd like to see some documentation. He asks if they can add an image of the TECP to be sure.

If the collision damaged the rasp, there would be no ice and who knows what kind of chaos would ensue. The myriad options are too hard for my brain to parse.

"We will unsafe the RAC, but not the RA or TEGA," Chris says.

Ray tells the team there's a lot of opportunities for other groups to jump in with any activities that are ready to go. He calls it opportunistic science.

"Please honor the five-meter rule and proceed lightly," he says, trying to defend the impending onslaught of the SPI I.

Chris Shinohara posts meeting times for the three anomaly resolution groups on a whiteboard.

"We might need to add a partition to the penalty box," Joel says.

"Don't worry," Ashitey says to me. "The mission will speed up after they meet the minimum success requirement. Then they will be less conservative. A lot of the flight rules can be relaxed and we can get on with it."

NILTON RENNO TAKES HIS PLACE IN THE FRONT OF THE ROOM AND smiles. He looks excited for his science talk. The first few slides of his Powerpoint presentation are of the same Snow Queen image that we talked about this evening when the shift started.

"The thrusters melted bits of the permafrost and splashed liquid droplets onto the lander legs," Nilton says, advancing his slide presentation. "These droplets form an exotic kind of brine." This brine is similar to antifreeze that is so salty, its freezing point has been dramatically lowered.

There are measurements, arrows, and circles pointing to various features of the images. It's like any of the other EOS talks, except for the palpable discomfort in the room. Scientists squirm in their seats every time Nilton says "next" and a new slide appears on the projector screen.

"The droplets remain in liquid form—" Nilton starts to say.

There's a distinct murmur. Groans. Coughs. Either this is the worst presentation in the history of EOS or Nilton is up to some scientific shenanigans. The crowd is going to start throwing fruit any minute.

"I disagree," Mike Hecht leaps from his seat. He can't contain himself one minute longer. "Why wouldn't it just be frost?" Hecht asks incredulously.

"It could be a lot of things. There could be any number of explanations." Hecht is beside himself, and the other scientists support his objection with "yeps" and "I agrees."

We should pause for a moment to locate the exits in case the fomenting mob gets violent, and figure out why everyone is getting upset.

To sum up: Nilton Renno calmly asserted that he made a major Mars discovery. He has evidence there's liquid water on Mars. You should know that there is *not* supposed to be *any* liquid water on the surface of Mars. That might be a strange point that you're looking for ice, made of water, but floored to hear someone say that there is liquid water. That's because the low-pressure conditions just make it impossible for there to be any liquid water. Any liquid water would sublimate, disappearing into the atmosphere. There might have been water a long time ago on ancient Mars and we're here to understand that. But right now, our Mars understanding only allows for water to be solid (as ice) or gaseous (as vapor). Full stop. Nilton is trying to tell us otherwise. Any over-achieving fifth-grader with a minor planetary obsession can tell you that it's not possible. Just ask Lucas, Nilton's son.

This mission was going to prove there is ice on Mars and hopefully some organic material. This is a lot bigger than any of that. If there's *liquid* water on Mars, planetary science just took a big leap. It will radically change our view of Mars forever.

Mike Hecht isn't buying it, and neither is anyone else in the room. Where is the data from the wet chemistry experiments or TEGA supporting this? How could he possibly make such a wild claim just by looking at a few images? If this is a joke, Mike Hecht doesn't find it funny.

Doug Ming, already skeptical of any big claims made at the end of sol science meetings, says the droplets could easily be explained some other way. For instance, the lander legs are the coldest local spot and frost is forming from vapor in the atmosphere. And they're certainly not liquid water.

"It would drip off if it was liquid!" Doug Ming says, nonplussed. Maybe this is another one of Nilton's stunts to rile people up. Objections come from all quarters. Peter asks the fomenting mob for quiet, so Nilton can at least finish his presentation.

Nilton says he has evidence. But there are too many objections to hear it.

"Frost on the lander legs would explain why the nodes there appear to change," Mike Hecht challenges Nilton again.

"I assure you that *these* will support the claim," Nilton says, pointing to a list of chemical compounds. He then presents some possible candidates of the chemical reactions that might be responsible for creating this brine. He says he knows he still has a lot of work to do on this, but he thinks we could use the TECP and some other measurements to help confirm his hypothesis. There is more grumbling.

"If this is soooo common, why aren't we sitting over a big puddle with a minus 60-degree freezing point?" Mike Hecht says, practically choking on his own rage.

"I can tell you for sure there aren't salts," Doug Ming says.

"WCL adds water and doesn't show salts," Diane Blaney, a soil science co-investigator, says. She agrees with the anonymous heckler in the back. There's a lot of hostility. Soon it's just an unruly science mob berating Nilton. They pummel him with science orthodoxy. Nilton tries to defend himself, but the attacks fly fast and furious and he can't keep up. Nilton takes a defensive posture.

"This event might be isolated because these are areas we might have melted in landing and caused the brine to splash up on the lander legs," he responds to Hecht. But the moment passed. He raises his arms to protect his face and gets nailed with another body shot.

Nilton is kind of shell-shocked from the beating. He just stands there absorbing each blow.

"Would you accept *any* scientific evidence that would suggest you're wrong?" someone shouts.

"We could just calculate the growth rates of the nodules to see if they work in the proposed model," Nilton says weakly. No one is listening. He gives up.

"Thanks for the feedback," Nilton says. "That's it." He takes his seat.

CHAPTER SEVENTEEN
AN ENEMY AMONG US

SOL 48

"WE HAD AN INTERESTING DAY YESTERDAY," DOUG MING tells the team at kickoff. "We had our first accident on Mars."

I don't think he is referring to Nilton's talk.

"We may decide that if the images come down and everything is hunky-dory, we can just move on." Ming makes no mention of the incident at yesterday's EOS. It's back to the business of planning the mission.

"Yestersol, three instruments were safed. Now we're in recovery mode," Ming says. "We'd like to begin more trenching and cleaning so we can do our rasping and focus the next few sols on demonstrating to NASA that we can do these things."

Today, if we do not feel it's safe to move ahead with RA, then we will have to move ahead with other instruments.

"It's a MECA play day!" Suzanne Young says. But before she can get too excited, Doug tells everyone there's even a backup—pictures and weather measurements—if MECA doesn't work out, if things are really not good.

He dismisses the team.

NILTON RENNO PULLS ME ASIDE.

"I'm more convinced now than ever that there's liquid water on Mars!" he says. Nilton spent the day working out some of the issues in his initial presentation, which was met with such hostility earlier. Apparently there was some relevant criticism bundled in all that yelling. Nilton spent the evening bolstering his argument, and now he thinks he has it.

"The geologists feel threatened," he says. Some nameless scientist spoke with him privately after the talk. He or she was not happy. Nilton tells me they told him that they've been studying this for twenty years and now you're coming up with something pretty far out there. Nilton started a turf war. They probably won't fight it out with protractors in the parking lot after shift, but there's a lot of animosity brewing for a team that still has to spend the next month under one roof.

"It's making them uncomfortable," he says. Nilton is either awesomely brilliant and they really feel threatened, or he's totally lost the plot.

"It's frustrating and not fair. The EOS is supposed to be a forum for open ideas and dialogue. Hecht throws out a lot of ideas that are very preliminary. So it's not fair to attack other new ideas," Nilton says. "There were people in the room who thought it was an interesting talk, but they were afraid to say anything because of how everyone reacted." Fear is not an emotion that you usually associate with science talks. Even though it's not healthy, it was very palpable that day. The anti-liquid folks wanted to tear Nilton to shreds. Now that he had a chance to re-group and rally some support, he says "Bring it."

THERE'S A WOMAN TALKING TO RAY ARVIDSON. SHE SAT THROUGH kickoff with a recorder and notebook and they talk breezily. Ray and this woman seem to be friends. He introduces her to a few of the JPL folk.

"Why don't I take you to see the lander and meet the PIT crew?" Ray asks. And they head off. What's this all about? Competition.

A few days ago, Miles Smith, a JPL engineer of no relation to Peter, asked me if there would be any trouble with this woman coming to write about the mission for a textbook. Smith had volunteered to be her escort. Sara Hammond was okay with her coming for two days, and I guess he thinks I might have some say in these matters. Still, I thought it was really kind of him to mention it.

"No problem," I said. It's fine. I don't feel threatened. He didn't even need to say anything. The EPO office approves the journalists.

Still, who knew it was going to be like this? Miles Smith's friend already knew everyone. Ray already gave her a personal tour of the SOC. I'm lucky if he slows his gait when I try to track him down for a question.

Miles said he wanted to make sure I knew she wasn't there to scoop me. But I didn't realize she already knew everyone. What if she's better than me? (Mars Lag symptom 62: Extreme jealousy and self-loathing.)

Chris Shinohara has everyone seated and ready to go for midpoint a minute before its scheduled start. It's a notable accomplishment. He wants to unsafe the instruments and get back to digging. We barrel through. The RA team is good to go. Joel, the fearless phase lead, RA team, and the science lead look at the robot arm data and agree on a plan to un-safe and get back to scraping.

The downlink engineers give their respective instrument reports:

"RAC is healthy."

"LIDAR is ready to proceed."

"TEGA is still safe but eager to get back in the game."

"The AFM did not optimize."

The SSI team presents 3D images of the RA smashup. Everyone puts on their blue and red glasses.

"If you look closely, you can see the bow wave," Ashitey says pointing out where the rock moved. "We see no reason why we can't pull ourselves out of this mess quickly." We break for the mission managers to tag up and agree to the plan.

While we're waiting, the mystery woman sits down with Carol Stoker. Geez, she's ambitious. Carol chats breezily with my new nemesis. From what I can gather, she is writing a book about astrobiology. Well, at least that's what she said, when I briefly interrogated her about what her intentions are with my mission.

I have a lot of strange territorial feelings that I still need to work through. Damn this Mars lag. My nemesis hasn't spent the last two years trying to persuade Peter to let her come on the mission. She makes a phone call and her friend *Miles* is all la-di-da come tour with me around the SOC. Well, this is my story—err—our story.

"This is not a search for life," Carol Stoker says to Ms. Nemesis. Carol is responsible for understanding habitability of Mars. That often gets mixed up with the actual ability for people to live there. Habitability just means the potential for *anything* to live there. So it's a good science term but a confusing regular-people term.

"The press wants us to look for life. I'd like to look for life, but that's not our directive."

Carol explains. I wonder if it's unethical to put my nemesis's quotes in my book. Nah, this is an open and sharing space. Any words said here belong to me—err—the mission. That's probably the best approach to transparency.

Oh geez, what's happening to me!? I should go home and sleep. Of course, that's probably exactly what *she* wants. I don't let her out of my sight.

AT MIDPOINT II, SUZANNE YOUNG TAKES THE MICROPHONE AND SAYS: "Welcome to sol 49 planning, everyone."

"We're sorry your call did not go through," sounds the machine on the conference phone.

Everyone laughs like mad. I think I see the *other* writer write that down. Damn. She's taking my best comic material. *Echh, let her have it.*

The core activities for the day are unsafe the RA, get back to trenching, and work on verifying the rasping blocks.

Bob Denise is the mission manager for shift II. He grumbles a lot as Suzanne reads the entries in the plan.

"This is too much work." Denise says.

"Does the RA team have sufficient resources to deliver these activities?" Bob Denise asks Bob Bonitz. Bob Denise clearly doesn't think they can finish in time.

"We can do it," Bonitz says. "We have 80.00001% confidence." And that's just enough to move ahead with the plan.

TODAY'S EOS IS NOTHING LIKE YESTERSOL'S SCIENCE SMACKDOWN. Leslie Tamppari gives a very high-minded practical talk about publishing your Phoenix research in a very special issue of the venerable scientific journal *Science*. In the "publish or perish" world of academia, this is important stuff. For some, it's everything.

"Each paper must contain science-worthy data, results, and conclusions," Leslie says. Ray Arvidson offers a tale of caution.

"Our special MRO edition was a disaster. On the CRISM instrument, the P.I. paper was rejected while the CO-I's was accepted. The CO-I then had to withdraw. Then the HiRISE paper was rejected outright."

"Maybe *Science* should not be our first choice then," someone suggests.

Some team members feel a tremendous pressure to publish quickly and prestigiously; it's *Nature* or *Science* or bust. Others feel like they don't have enough time to deliver meaningful results and would rather take their time. This is an old debate because of that pesky data-sharing rule. Data will only be private for six months after the primary mission ends. Then anyone can jump in and use the research to do their own analysis and publishing. Some of the more concerned team members worry that if results leak, other scientists could hijack the work and publish, scooping them on their career-making research project.

"It happened on MER," Mike Hecht says, talking about the last Mars Expedition Rover mission. "They scooped the jarosite results."

Ray Arvidson shakes his head.

"That's just cheesy," Ray says. I don't know what happened on MER; but with competition lurking, I know how they feel.

Before the end of EOS, I ask Carol Stoker about Nilton's liquid water theory. She shrugs.

"I wish he'd drop it for now," she says. "He's hurting our case for digging in D-G," she says quickly and goes back to work. She seemed much chattier when my nemesis interviewed her. Damn.

CHAPTER EIGHTEEN

NILTON'S NODULES (ROUND II)

SOL 49

NILTON RENNO LOOKS UP AND SMILES.

"I have a smoking gun," he says. He's sitting at his computer; it's still an hour or so before shift I starts.

"Morten made me an image. He subtracted pairs of images to show the difference in the nodules," Nilton says. The images are from sols 8, 31, and 44. There's a lot of stuff moving around. Some of the barnacles grow; other blobs fall away.

"This will show that the water vapor disappeared and the liquid salts are growing. This is evidence of the brines . . . evidence that liquid water can flow on Mars!" Nilton says. Wow.

I know just enough to know that this is big. I'm here for some kind of "Eureka." I'm not quite sure how to comport myself at such a moment of discovery. End-zone dance? I have yet to see anyone spike their laptops, so maybe that's not appropriate. There's not a lot of training for these sorts of things at space camp. I express my enthusiasm with a hearty handshake and a genuine exclamatory.

"That's amazing!" I say. I also want Nilton to fill me in on the mud-slinging that's happening behind the scenes. It's slightly easier for me to digest. There are lots of emails flying around that continue the assault on "Renno the Heretic."

Nilton says he isn't too fussed by the naysayers. Potentially making the biggest Mars discovery since the discovery of the planet helps numb the pain. It's a sweet salve. Sure, he's a little annoyed that people aren't more supportive, but he has his allies. And today Nilton is going to do a follow-up presentation.

"If I can make a thermodynamic model, the case will be bullet-proof," Nilton says. "I sense there will be a lot of opposition." There's no hint of irony. I personally hope no one throws anything. I need to just get there early; news is spreading quickly.

I SIT NEXT TO MORTEN AND WAIT FOR KICKOFF TO START.

"I would co-sign a paper on liquid water," Morten Madsen tells me. He thinks Nilton has a beautiful idea and he would love nothing more than to support it.

"I'm not 100% sure he's right just yet, but the evidence from the image should be looked at." That's noble. He's working to help Nilton solidify his idea.

"I can't make it to today's EOS. Do you think you can give me a report?" Morten asks. I would love nothing more.

Kickoff begins.

"Today is a big day," Doug Ming says again. He should consider a trademark on this line, seeing as he uses it every day. And it's true. Still, I fear for the day that's small.

"We're hoping to do a rasp and scrape test! And we're looking to deliver to TEGA by sol 56. There are a lot of activities in the plan today. Almost every instrument is involved," Doug says. It's all hands on deck.

He turns things over to the instrument teams.

"We successfully extracted ourselves," Ashitey tells the team. He is happy to report the RA successfully unsafed and got to work scraping. Most of the scrapes went very well, he assures us. The RA team says they really pushed their limits yesterday by scraping the hard stuff with all their might. The idea was to examine what mix of scraping and rasping is optimal for the big ice delivery. Then something went awry.

"Then we exceeded our resources. And we believe that's what happened," Ashitey says, telling us that the RA is safed again.

"RA had 54 minutes allotted in the plan to scrape. When it got about three to four minutes away from that time limit, it recognized that it would not finish its task. When it knows it can't finish a task, it starts to park itself and then safes to prevent other commands from executing," he explains.

"It did what it was supposed to," Ashitey says. We all feel proud of our tough little arm even though he's safed again. Then I remember a little trick Ashitey once told me.

"You have to say something nice or fun to put the scientists at ease. Don't ever just say the instrument has 'safed.' Too often they just hear 'safe' and it gets them all worried. Then you have to explain yourself. You don't want them worried. There's no need and they have other science things to worry about. If they feel good about what's happened, they'll let go." He told me this once when he didn't have time to explain what was going wrong with the RA. Before he walked away, he added, "Now you probably know more than the scientists." I felt really great, for a second, before I realized he taught me this trick and then used it on me. He's good.

TEGA is also in safe mode, and its engineers report they were able to replicate the heating issues on the engineering model in the PIT (where Phoenix II lives). That's not as compelling as the RA news.

"Today we need to unsafe TEGA and figure out if we need to continue digging and scraping or none of the above," Doug says. We break.

ECHH. THE WRITER GIRL IS BACK . . . AGAIN. I HAD A TERRIBLE DREAM where I threw the scoop at her. That's so hostile. I was fine with her sitting in kickoff, even talking to Carol, but now she's waiting to talk to Nilton. He's my source! Does she have to talk to everyone?

I'm so distracted by petty jealousy, I hardly notice all the scrambling going on around me. Doug Ming calls the group together.

"We have to redo the plan," he says. "There is a lot of scrambling because I screwed up. There was an APID missed. And now we don't have the images we need to go ahead with the plan." A sign outside the SSI office says "5 days with no lost images." But there won't be a sixth. The SSI team created a new 3D map of the trench and it's lost in space. The image is key for the RA team to better understand the scrape area. No one is sure the scrape test is working, because the images are inconclusive. Without the image, the RA team doesn't want to move ahead. Doug says he downgraded the APID accidentally, in a burst of scrubbing to get the plan cleaned up. Now the image they wait on won't come back to Earth until the next communications pass. It's too late to be useful.

There is a new plan coming together. This time around, Doug goes through the APIDs very carefully.

"Several errors could probably be blamed on fatigue already," Joel says. "It's part of the reason we want to get back to Earth time." Rich Volpe from the RA team just warned me about this sort of thing. Since it's getting later in the mission, there's a feeling of urgency mounting. Team members will push to maximize the number of activities in a plan. That's the wrong approach. It results in sloppy errors and accidents from both humans and machines trying to do too much too quickly.

"The optimal plan has fewer than the maximum number of activities," Rich tells me. Yes, they say it like that, because whenever they offer dire warnings, it's in their awesome engineer-y syntax.

LET'S GET READY TO RUMMMBLE. NILTON RENNO VERSUS SCIENTISTS: round two. Even a few of the shift II engineers sneak away from their sequencing meeting to attend. It's packed.

Miles Smith's writer friend floats in and takes a seat that one of the scientists reserved for her. Geez. There aren't usually any visitors in the EOS, but I guess she's a VIP. She's just getting comfortable when—

"Is everyone aware there's a journalist present?" Mike Hecht asks. Uh oh. "I'm sorry to do this." Hecht apologizes to Ms. Nemesis. Apparently he knows her too.

"I'm not comfortable with her in the room. I propose we discuss if this is a good idea, before we continue with any, possibly contentious, EOS talks," Mike says.

Nilton smiles. Vigorous debate ensues. The poor woman is sitting right there. It's awkward, and I'm really glad to escape people's ire this time. Some say there should be openness. But most everyone thinks that if they are going to express half-baked ideas, the press should be excluded until there is a concrete plan or some conclusion. So far, this has been a closed forum.

"I don't want to have to watch what I say and I don't want anyone to print anything that would embarrass the team," Mike Hecht says. I wonder if he also meant me in that line. I stop taking notes.

"So you'd like her to leave?" Sara asks.

Yes, they would. Now it's even more awkward, and she seems miffed. She pouts as she grabs her things. When she gets up to finally go, she looks at me for some kind of defense. I slump low in my chair and search for an aphorism that lets me feel less guilty about being a coward. *Discretion is the better half of valor.* That works.

Now that she's gone and we're free of any "intruders," we can get started.

"When the lander touched down, the surface temperature directly under the lander reached approximately 500 Kelvin." Nilton starts his second groundbreaking and contentious liquid-water talk with landing day. He builds up to the events that formed the nodules. Landing caused mud and salt to splash all over the place.

"Then the top millimeter of material cooled quickly," he says. This is the starting point for his argument about thermodynamics. It gets a little hairy here. But he says the thermodynamic properties of saline solution melting cycles produce eutectic solutions. This time I've come armed with a month's worth of chats with Nilton. I hold tight to my tenuous grasp. Here's how I understand it: We have a salty brine, like pickle juice, but this salty liquid is a Mars sludge. It formed through a process of pure water freezing out of a salt solution. Each time the temperature changes, more pure water freezes and the saltier part concentrates—the more water that leaves, the higher the salt concentration. Since salt lowers the freezing temperature, this lowers the melting point of the solution. If this process is iterative, heating and cooling more and more, the pure water will slowly freeze out of the solution and soon you have some kind of super antifreeze, our Martian pickle brine. Instead of regular table salt, we now have something a bit more exotic.

"That's what we see on the lander legs on Mars," Nilton says.

This is an idea Nilton toyed with for a while. Now it clicked. Nilton directs our attention to an image of the lander leg. The crowd is polite and listens quietly.

"The larger particle is growing and the small particle is shrinking. In general, growth is a function of size," Nilton says.

"You're not accounting for thermal inertia," Mike Hecht says, breaking the silence.

Nilton ignores the comment, but his face stains and his cheeks tighten.

"This should demonstrate the principles I have outlined," Nilton says and then walks the team through a series of equations.

"Thermal inertia is not accounted for!" Mike Hecht objects after Nilton finishes his mathematical rundown.

"I'm not sure I agree with that," Bill Boynton says to Mike Hecht.

"Maybe it's hydrazine," Mike says. Hydrazine is the rocket fuel the lander used in its propulsion system. There was some concern it might contaminate the landing site if it leaked. Nilton knows the ins and outs of the hydrazine situation and is prepared for the questions. His group conducted several experiments before launch. Suddenly all of Nilton's extra-curricular experiments make sense. Nilton ignores the side debate brewing. His next slide reads "How can we test this?"

There's a bullet-pointed list of ways to test this hypothesis.

- Estimate the atmospheric humidity near the leg
- TECP measurement nearby
- Ice temp
- Estimate leg temp
- Calculate droplet growth rate and compare calculation w/ measurements

"Thank you," Nilton says in conclusion and sits down. There's no fight this time.

The talk is over, and the engineers leave before the short-term planning meeting begins. The meeting doesn't last long before Joel Krajewski comes in with an update. TEGA is pulled from the sol's plan. Their pre-heating fix still doesn't work. They need another two sols to update flight software so work can continue.

IN SMALL GROUPS, MEMBERS OF SHIFT I TURN OFF THEIR COMPUTERS and leave. There's a going-away party for one of the young scientists who works on the Atomic Force Microscope. She's getting married and won't be back before the end of the mission.

Bill Boynton offers me a lift to the store so we can pick up some celebratory wine. What are you supposed to drink at 8:00 a.m.: red or white? This is when my Mars watch comes in handy. It's just after five on Mars. I go white. Boynton gets red.

I ask him about the wiring problems with TEGA that still continue to plague us. This is probably why they don't invite me to too many parties.

"The wiring was done poorly," Bill tells me again with a sigh. These are problems that were baked in when Phoenix inherited instruments from the Polar Lander. TEGA has lots of wires packed into a small space. The designers saved space by sharing wires across circuits. They got a lot of savings by using the same circuits for all eight sets of TEGA doors. It seemed like a good idea at the time; but now that one has problems, they all share in the mess. Any yet, Bill doesn't seem bitter or resentful. Just the opposite.

"I'm happy with the data we have," he says.

WE'RE NICELY BUZZED FROM OUR MORNING COCKTAILS BY 8:45 A.M. We hang out, joke around, and share tips for sticking with Mars time.

"You have to ignore all clocks," Sam Kounaves says, although he admits he just fell off the wagon this week after a quick trip back east.

Heather Enos arrives late. She pretends to kick me out again. But I'm prepared this time.

"If you're willing to stick with us this long and take our abuse, you can be part of the team," she says. I feel warm and fuzzy.

WHILE WE ENJOY OUR COCKTAILS, PHOENIX WORKS. THE RASPING TEST should be happening right now. Phoenix firmly plants the back of its scoop on a flat section of the hard material. It's got to get the position just right; the area the science team chose is no bigger than a large fist. When it finds the proper position, Phoenix presses the load plate on the back of the scoop into the bottom of the trench. Applying firm pressure, the rasp slowly begins to turn, pushing its way into the cemented ice. After some careful drilling, Phoenix lifts its arm to examine the site. Phoenix snaps a photo. It worked. Now for the tricky bit. Phoenix has to articulate its wrist front, back, and front, so the rasp-tailings collected in the special chamber flop over the barrier and into the back of the scoop. If it works, Phoenix will send its documentation home for us to evaluate and then get some well-deserved robot rest.

CHAPTER NINETEEN
FEEL MY RASP

SOL 52

I SLEPT THROUGH MY FIRST ALARM, MY SECOND ALARM, AND THEN the whole shift. Seventeen hours of sleep might be a new record. I guess I was tired. It's probably not normal to lose an entire day. So long, sol 51, how we shall miss ye.

The SOC is oddly quiet three minutes before kickoff. With thirty seconds to kickoff, there's a mad rush of bed-headed scientists filing reports and filling seats. I guess I'm not the only one who's tired.

Vicky Hipkin is the sci-lead today. She looks over the first rasp and pronounces it a great effort. Good work, Phoenix. Soon we'll see the results of yesterday's expanded scrape test, and then we'll complete the scrape rasp combo for tomorrow's big activity. A full dress rehearsal for the TEGA delivery. This is it: twenty or so coordinated activities to prep, drill, scrape, and then practice the delivery. These coordinated events put Julia Bell's sequencing engine achievement on glorious display. The interoperation is denoted in the plan as "The Monster." If they can nail this, NASA will give Phoenix permission to proceed with the real delivery attempt. It's a short day, with laser focus on the rasp and opening the TEGA doors. We could have ice and be outta this mess in a couple sols.

Shift lead is Miles Smith.

"There is a very aggressive tactical timeline. And there is only one uplink opportunity," Miles says. That means if things are running behind in terms of the data upload, they can't switch satellite relays and extend their day. There's just one shot to make uplink. If today's plan is successful and we can deliver "ice" to NASA, we should have our mission back in a couple sols. Miles pleads with the team to work efficiently to make this uplink and stay on track. The engineering team hasn't missed one yet, and he doesn't want the first to happen on his watch.

There's just one problem. TEGA might need to update its flight software again before proceeding. It seems there's a file missing.

"Is the file needed for the TEGA door-open activity?" Miles asks.

"Open door should be okay, but 'Check-Out 1' does need the file. So there's 1-in-700 chance when the EGA turns on, it safes. There probably should be a lien," Dave Hamara says. Mission managers put a lien on a plan when they need to move forward but have an unsolved issue. If they don't solve it before uplink, they know to pull it from the plan.

What could possibly go wrong?

"Please look at yestersol's scraping activities carefully to be sure it's a well-prepared surface. This might be our one go," Vicky says, imploring the team to look closely at the new data when it comes in so we can optimize today's plan. She too wants us to put this episode behind us and move on.

We start to work through the day's activities, examining each for possible conflicts or problems. They call this "scrubbing" the plan. These plans can be messy. This plan is no exception. There are conflicts, inconsistencies, and activities that still need to go through the validation and verification process. This collective editing process helps the SPI I catch errors before the plan gets to the approval stage when engineers can cut out anything they don't like. Some notice that the RAC and SSI heaters are scheduled to run at the same time.

"Can we do that?" an engineer asks.

Then as downlink starts, there are more delays. Neither TEGA nor the RA unsafed. RA timed out yesterday after hitting hard material. And we won't have the scrape test from yesterday to help us today.

"Isn't there a time-out block?" one of the systems engineers asks.

It's not validated yet. We can't use it.

For the engineering fans out there wondering why TEGA and the RA didn't unsafe, here's what happened: no one told Phoenix to tell the RA and TEGA to unsafe. We solved their underlying problems and it was time to turn back on; they just didn't know. That's the simple part. The reason for not telling them is more complicated.

The "science master" is a file that tells the spacecraft when to wake up and go to sleep each day. Systems engineers carefully prepare these schedules in advance to make sure that this document is in sync with another important engineering file called the background master. It's the background master's job to tell Phoenix when the relay orbiters are in range. The background master then nags Phoenix to phone home. It's mothering software. They uploaded an updated version at the last minute, but no one included the unsafe commands attached to it. Since Phoenix looked to this file to see what to do first, it didn't know to unsafe the RA and TEGA before it tried to use them. Robots are so literal.

The modicum of excitement present when we started this meeting bleeds from the room. It doesn't seem like we will get our ice. New problems with the hastily constructed plan come to light. With a flight rule violation here, a lien against an instrument there, and a key activity forgotten, it seems like we need to temper our excitement.

Things feel a little chaotic. Midpoint II is delayed. Twice. The tactical timeline starts to slip away.

When we finally get started, Vicky hands things over to the SPI I. Things should move quickly now. Today's SPI I is Mike McCurdy. He built the planning software, Phoenix Science Interface (PSI), that the SPI I uses to schedule Phoenix's day. Yes, the SPI's use PSI. NASA acronym humor. Good thing Mike is on duty. He's our best hope to get back on track. We need someone to work quickly; it's gonna be down to the wire. But before Mike can say a thing, PSI crashes, then the computer goes down. This is awkward.

It becomes clear that there's too much work and too few tired engineers to rush this TEGA delivery through today.

"Perhaps we should consider eliminating the test?" Miles Smith asks the shift II lead and the shift II mission manager.

"You're going risk pushing the TEGA delivery back by days if you don't do it now," Heather Enos says with exasperation.

Joel Krajewski steps in.

A few weeks ago, when I was about to leave the SOC after a 16-hour day, I passed Joel Krajewski in the hall. The SOC was still full of scientists and engineers. I said I didn't know how they do it so many nights in a row.

"Crushing guilt," he said. People stay all night to make sure things are perfect because the guilt that someone would feel from a mistake that ruins the next day destroys them from the inside.

I guess fifty days is a maximum for lander guilt. There are still many late nights, but with so many new problems it's just not possible to stay a few extra hours and triple-check your work. The guilt-transport-system slows. Guilt, excitement, or anything but a deep passion for sleep are muted by this point in the mission.

It's becoming clear that at least from a psychological perspective, Nilton Renno and Carol Stoker were right. We should have scooped ice from Dodo-Goldilocks weeks ago and moved on.

Joel tries to clarify what's at stake to head off a fight. A good mission manager lays out the consequences for taking each path and then lets the stakeholders decide.

"There are two critical paths running in parallel, RA and TEGA," Joel says. Both need attention in order to move ahead in a timely fashion. He explains that if rasping works well, we need to be ready with TEGA, which requires a one-day gap between opening the door and delivery. Dropping the monster activity will cause a delay. Not getting it done properly will too.

Everyone agrees: let's push ahead with the plan. Team members vote with their 80% confidence in the plan.

THE 50S ARE NOT A GOOD TIME FOR THE MISSION. EACH DAY BEGINS with a new hope that's quickly dashed. The test on sol 53 is a moderate success. They manage to open the TEGA doors, only to discover a problem with a frozen valve. The valve is fouling up the internal pressure in TEGA. Then a problem with the load plate at the back of the rasp pushes back the next drilling exercise. Pre-heating and validating code goes slower than expected. Waiting is not interesting.

CHAPTER TWENTY

MARTIAN COLDS

SOL 57

O N THE FRONT PAGE OF TODAY'S *TUCSON CITIZEN* APPEARS A picture of George W. Bush waving to adoring fans in Tucson, Arizona. The President of the United States comes to the desert; he probably wants to see firsthand the amazing discovery made by Phoenix. Nope, he doesn't stop by Mission Control. Doesn't he know there's a Mars mission happening here? Maybe he wasn't invited. Maybe he doesn't care.

RA team stayed up late looking at yesterday's rasp test. They seem frustrated. I even think I hear one of them curse. No, he said "shoot." Sorry for the misunderstanding. Something has changed. Where is the happy-go-lucky, get-the-sample-at-all-costs team I know and love? It seems that the RA team has a visitor: contempt.

"They still don't get just how hard this arm is to operate," angry RA engineer #1 says. They're worried I'm going to write about their anger, so I'll protect the identities of the cranky.

"This is not like the arm on MER! How many times can you say it!? This arm is two times longer and doesn't have contact sensors. And the DEMS are so noisy. What are we supposed to do with those? You try to measure distances in the images and run them through the compliance models to understand the flex," overwhelmed RA engineer #2 says.

Having never used a compliance model, not once in my entire life, it's hard to empathize. Still, it's good to vent their frustration. They're frustrated because they have to precisely position the load plate at the back of the scoop. They spend a lot of time on this placement. There's a lot of pressure to make these precise movements, quickly and without all the information they need. Every day they don't deliver ice just makes the feelings of frustration grow.

And I'm not the only one who noticed the tension. The mission managers mandated a pre-delivery down day for the RA to make their final calculations. The day before delivery, the RA team will spend their day calculating and coding only. And by official order are excused from any other duties.

JOEL KRAJEWSKI CALLED IN SICK YESTERDAY. PETER HAS A COLD. And there are ten or so other sneezers blasting germs into the SOC atmosphere. Mars lag is slowly killing us all. It's time to consult Edna Fiedler from the counter-fatigue group.

"I'm glad to see you," Edna says.

Before I can start complaining, she says there's something she'd like to speak to me about.

Me?

"Yes."

Then she waits for me to talk. This is an awkward exchange. It seems my basic social skills are breaking down too. We both just stand there.

"Don't you have some questions about how you're feeling?" she asks.

It's like she can read my mind. I have questions about how *everyone* feels. She says she can't tell me about the others because that's confidential, but I can ask her about myself. And then, we agree, it's all about me anyway.

"There are probably changes in your T-cell count happening," she says. Oh dear. My immune system is not working efficiently. I knew it.

"There are possible pulmonary problems and the liver too," she adds. "Your endocrine system is slugging along. The function hasn't stopped, but it's sluggish. You're not moving hormones." All this is due to the slow but constantly debilitating sol schedule.

"So what's the good news?" I ask.

They're really learning a lot for future missions. So Edna is pleased. Now that I know what's wrong with me and why I feel this way, it's her turn.

"Have you ever heard of astrosociology?" Edna asks. I haven't. Great, one more thing I'm supposed to know but I don't.

"'It's the sociology of space travel and other planets," she says. "We're having a conference and we'd like you to speak about your time here. It takes place in Huntsville, Alabama." I play coy.

"Oh, that sounds interesting," I say. But really it's the most awesome thing ever. Suddenly my symptoms don't feel so overwhelming. Fiedler says she'll email me the details right away. How can I resist? She goes to lunch. I Google astrosociology and start checking airfare to Alabama. I wonder what I'd talk about. Maybe the bags of pee everyone always carries around Mission Control would make for a nice opener.

IT'S MORNING IN TUCSON, EARLY EVENING ON MARS. THE START OF SHIFT I nearly syncs with a regular Earth day. For a few glorious days we'll come into the SOC at seven, eight, nine, and then ten o'clock in the morning. Even if our bodies are getting worse, there's a huge psychological benefit to waking with the sun. The last week of coming in at three, four, and five in the morning was especially demotivating. And that's on top of the continued hiccups with TEGA and the other elements of the dig.

It's nearly time for midpoint. Scientists mill about before taking their seats. Bill Boynton grabs the microphone.

"Everyone needs a balloon for the meeting," Bill says.

Heather comes out of the TEGA office hugging an armload of balloons.

"Please, you must have a balloon for midpoint," Bill repeats. And he goes to help Heather pass them out.

Three minutes before midpoint begins, everyone has a balloon. The balloons come with smiles and a little confusion. We look at each other and wonder what wondrous discovery warrants this great balloon giveaway. Anything to get us through these doldrums is appreciated. There has not been a lot of good news in the last few days, but balloons are a universal indicator something great is going to happen.

"We're having a prickly-ass contest," Heather says. "Everyone will sit on their balloon and whoever pops the balloon first, wins. Get ready . . . Go!"

There's no time for questions. An overzealous engineer goes splat when his balloon rejects his overture. We race to pop our balloons. Popping and laughing ensues.

But that's it. There's no news. Just a random balloon-popping contest to keep us from bursting ourselves. Well timed. We all win.

The fun is over. Doug Ming starts midpoint.

It's been a slow and monotonous few days. The most exciting event was a giant monsoon that collapsed the roof of the small conference room. This is where we hold the end-of-sol science and planning meetings. Not any more. Did NASA send the storm to put a little more pressure on us? I'm not at liberty to say.

TODAY, TEGA IS FINALLY UNSAFED. THE PROBLEM WITH THE VALVE IS under control. The pre-bake checkout looks good, and so does everything else. That's good news. All the big hitters from NASA and JPL are here. Ramon, Barry, and Dara are back. The President is somewhere in Tucson. And the instruments are nearly ready. Let's do this. We could get past this NASA imposition any day now.

"No data gaps or errors," reports a TEGA engineer. TEGA is ready to accept a sample. As soon as NASA says "Go," the TEGAns will open their doors to make way for glory.

Apart from the curmudgeonly RA engineers, things are going great at chez RA. They just tried their autonomous fault recovery. It's a scheme where the RA times out but does not go into safe mode. Fault recovery is a bit of code that lets the RA stare a problem in the face and then instead of backing down, it confronts the problem and gets back to work. This only works for a few situations, but it should save a lot of time now and potentially prevent some future slowdowns.

THE END-OF-SOL SCIENCE MEETING IS IN DOWNLINK. BEFORE WE GET started, Pat Woida comes in with a big smile.

"Someone made lunch for the whole team," he says. A local woman, impressed that a real space mission was happening in her Tucson neighborhood, decided to cook authentic Sonoran food for one hundred people. There are tamales, stewed meats, vegetables, and salads. A small woman with an enormous smile stands near the feast and smiles at the convoy of hungry engineers. I make myself a towering plate of corn-husk-encased goodness and stewed meats. I get so lost in my plate that I fail to ask her name or why she fed us. I'm a good eater but a terrible reporter. Thank you, mystery woman; it was delicious.

CHAPTER TWENTY-ONE
THERE IS NO TRY

SOL 59

"**T**ODAY IS A BIG DAY," DOUG MING SAYS, AGAIN, TO START kickoff. "We will do our 16 rasping holes and acquire the sample. Then deliver and TEGA low bake. There will be little else."

The plan will go something like this: 5:00 a.m., wake up. Preheat RA. Rasping. Scraping. Preheat TEGA. Deliver sample.

The downlink priorities are all rearranged. The table that controls them gets a special update—tested carefully—to ensure that the TEGA data comes down with the highest priority tag, its APID. The engineers moved it above some of the most critical engineering data.

The weather report doesn't bode well for us. There are storms on Mars! And for the first time, the team sees ice clouds brewing overhead.

"Wow, the weather will be changing!" Doug says. It's autumn on Mars.

Everything is double-checked. The storms pose no threat, but we need to be extra-careful today.

"Measure twice. Cut once," an engineer says, passing some colleagues rechecking the plan.

Matt Robinson feels good about the load plate test results.

"We're set to give it our best shot. But a lot of things have to go right. sixteen rasps give you two cc's of material," Matt says. "I sure hope that's enough." The TEGA team has reset light/dark parameters on the LED for the oven-full signal. They're gaming the system a bit to give them an oven-full with a smaller-than-usual quantity. Don't tell Ramon.

Midpoint arrives and it's time to decide "go" or "no go." The teams report on their general status. RAC is healthy and ready.

"TEGA is healthy and clean, but ready to get dirty. There is no evidence of an electrical short," a TEGA engineer says.

"I looked at the RAC images," Doug says. There are no problems. He's ready to declare it a "go."

All that is left for the SPI I to explain the plan to the shift II team. They're relatively happy. But there are still lingering concerns. Questions and answers come in short staccato.

"What happens if MRO fails?"

"The data is protected."

"Memory concerns?"

"Critical data will be written to flash . . . with plenty of margin."

There are opportunities to re-transmit. Everyone gets their questions answered. The core plan gets the APID "0": highest priority to be transmitted from MRO. What about everything else? There is nothing else.

"Please cross your fingers and toes," Doug says.

"And pray to the ice gods," Bill adds.

Meeting dismissed.

"THIS IS IT FOR ME," DARA SABAHI SAYS. HE CAME BACK TO HELP BRING TEGA into the proper delivery position and offer some guidance to the team. "I've done everything I can." Now he's leaving for good. He doesn't know when and if he'll return. Dara says he's not worried about the team getting ice.

"If it doesn't work and it's an engineering issue, then it can be addressed. If it doesn't work because of Mars uncertainty, we can learn something and try to work around it. Either way, we can address it. The only time we don't succeed is when we don't try," he says. It's tough to argue with his life-affirming view of the world.

"Remember, this is about managing many points of risk. You must balance them but you can't run from them," Dara says as he passes the baton to his team.

"I will," I think to myself. Maybe I'm internalizing.

"Our clients can't go to the mechanic if something goes wrong." Dara is partial to automotive engineering metaphors.

"We have a prototype that has to work perfectly, the first on the road," Dara says. "You can do that, but it involves managing a lot of risk. And you might imagine that it's a technical challenge. And that technical challenge is complicated. But on these big projects, it's not as complicated as cooperation."

It's rare that engineers play down technical aspects for those of basic human interactions which are vastly more difficult to plan and almost impossible to predict. I guess that what makes Dara a universally praised member of this team. He sees space problems in terms of human interactions.

"The little things matter. Mutual respect matters. It goes a long way if you want to be successful." With that, Dara says good-bye.

He put his heart and everything else he had into getting Phoenix on the ground and scooping ice. And now he's leaving. It's time for him to get away. Maybe for good.

"I'm going to get in my car and drive. No computers. No email. No radio. No people. No politics," he says. "You can try to contact me. Just don't count on a response," and then smiles sweetly.

Mike Mellon is back in the SOC after a two-day rest. He had to go home and check on his house. This trip back to the real world reminds him of everything he's neglected over the last year. He tries to put it out of his mind now and refocus.

"It takes a minute to catch up," he says. "There are new images scraping and rasping. It's a lot." He's happy now that there's time to think about science goals instead of just this one icy sample that's been the focus for so many sols.

"This will be a good sample," Mike says about the new RA acquisition. "Probably similar to the one we were going to get a month ago."

Geez, time flies. That whole drama was a month ago. It feels like days.

"If we'd known it was going to take this long, it would have made sense to do other science. There is a lot of work to do with the RA and many science goals. Now we spent a lot of time and resources rasping and scraping one trench," Mike says. "But it's hard to look back like that," he says. There was not a lot of time for second-guessing, but when pressed it's hard not to consider what might have been.

"I just don't know if we should have done the Dodo-Goldilocks sample. Remember, it's of unknown origin. We could have got it, but it might mislead the science community if we say this is a typical sample. This new sample is what we want. But it's just taking a long time." The downside of taking so much time is what was missed. And that's a high price to pay.

"We've neglected change monitoring and some of the other features, like Snow Queen. Now we have a 20-sol gap where all those cracks and interesting things occur. But when did that happen? Slowly? All at once?" Mike asks. They don't know. This may seem like a small thing, but those observations could be the basis of great discovery. "But NASA had other priorities," he says diplomatically.

"I suppose she [Carol] might have been right," Mike finally says, willing to concede that, in hindsight, quickly grabbing ice could have saved a lot of heartache and headaches. At the time, though, they did what they thought was right. They cannot predict the future. They can just do science: hypothesize, experiment, and repeat.

Now we look forward. Before Phoenix landed, there were only fuzzy images of the polygons. And Mike was desperate for a close-up look. Now he's here and there are close-ups and extreme close-ups, but that doesn't feel like enough. When every scoopful deeper into Mars reveals some new hidden treasure, it's hard to feel satisfied. No matter what situation you're in, on a mission like this you always want more. That's frustrating, but also exciting. It's what keeps us going.

"I work on HiRISE [the imager on MRO], and those images are great. The images are tantalizingly close to being good for looking at the permafrost. Now we're there and we can get centimeter resolution. That's even better. But it just gets you wondering what's next. You always want to know more. You just want to get out and look beyond the next polygon," Mike says.

This is what keeps him coming back, staying up all night and going until his body fails him. There's always some new piece that seems so close you can taste it, but it will take years of work to finally get there.

Mike is quick to point out that this isn't supposed to be easy.

"This is hard stuff to get. I was in Antarctica looking at the dry valley. That permafrost was hard. I could barely make a dent with my ice ax!" he says. Mike is a pretty big guy; swinging an ice ax at full force should do some damage. "Now for comparison, that was in ice cemented soil frozen at −5c. And we're trying to get ice that's been frozen at −90c. So it *should* be difficult!"

BEFORE SHIFT I ENDS, THERE'S THE MATTER OF LONG-TERM PLANNING to address. All the new science experiments are after the ninety-day primary mission. Will NASA extend the life of Phoenix? How much can we get done once the team moves to Earth time and the scientists return to their home institutions?

"We've reached the point where we're reluctant to push stuff to the right (on the plan calendar)," Aaron Zent says.

There are still seven TEGA cells and two wet chemistry cells in MECA. We start to run out of time. Zent estimates the earliest they could get through all their samples would now be sol 120. That's thirty days into an extended mission—if NASA funds one—and dangerously close to the start of Martian winter. Which means there might not even be enough sunlight left to power the more energy-intensive activities. Everyone looks to Joel Krajewski for some insight. He nods affirmatively. What can he say? We're running out of time.

"When Joel nods, people listen," Aaron says.

BILL BOYNTON PACKS UP HIS THINGS AND TURNS OFF HIS COMPUTER monitor. He stands up and looks around.

"The die has been cast," he says. We are about to cross the Martian Rubicon.

CHAPTER TWENTY-TWO

DON'T BE A RASP HOLE

SOL 60

VICKY HIPKIN IS THE SCIENCE LEAD. SHE GIVES THE SCIENCE day invocation, "Welcome to exciting sol 60. We've all got our fingers crossed." Vicky puts up a special sol quote on the overhead projector. It says:

"A dream is a wish the heart makes."

"It's from Cinderella," Heather Enos tells me. "I picked it. The sample is called Glass Slipper. And once the glass slipper fits, we will be free from our curse."

To get in the spirit, Heather wore a special pair of pink heels, her "princess" shoes. I guess there's a little room for superstition in the hard sciences.

EVERYONE GETS TO WORK. THERE'S ANTICIPATION AND EXCITEMENT in the air.

Katie Dunn is in her office. She monitors the data as it streams in.

"It's not looking good," she says and sighs.

What? That can't be right. All this buildup . . . the funny sample name . . . the pink shoes . . . and then we get the sample and all is well in the world. That's how the story goes.

"TEGA safed," she says looking closely at her monitor, face downcast.

It gets very quiet in the SOC.

"I THINK IT'S AN 'OVEN FULL' ISSUE," CHRIS SHINOHARA SAYS. THEY probably didn't get enough material. If TEGA thinks it doesn't have enough sample for a heating run, it goes into safe mode. They have to repeat the whole exercise.

"Where exactly in the process did it safe?" Barry Goldstein asks Chris Shinohara.

"Unclear," he says. "There looks to be some dirt at the bottom of the TA." The TA (Thermal Analyzer) is the oven part of TEGA.

Mark Lemmon tries to process the new images of the scoop and TEGA as they come down. He types furiously, putting his face close to the monitor.

"The scoop had loads of material in it," he says without looking up from his image-processing software. Mark doesn't think it's an issue of not having enough material. Maybe there's something else going on.

The team divides itself into two. Mark and Chris's group try to figure out what happened. Ray Arvidson heads a team to work on a new plan.

"If we want to retry on sol 62, we need to scrape on sol 61," he says. Of course, then they'll have to cut out the rest of the exploration digging that's been scheduled. Bob Bonitz and the RA team don't look so good. They are probably a few activities from catatonia. I fear Matt Robinson could pick up the water fountain, throw it through the blacked-out SOC windows, and run off into the desert at any moment.

Mark says there were at least three cc's of material in the scoop. That should have been more than enough for a delivery. Why is TEGA reading empty? Did it stick to the grate? Not again.

WHEN MORE COMPREHENSIVE ENGINEERING DATA FROM TEGA STARTS coming down, there's even more mystery. The LED pattern indicates that lots of fine material passed the LED, but it's only reading a few percent full. And yet, someone says there's another means of measuring and that signal says it's 75% full.

There's a lot of conflicting information flying around. But that's pretty typical when the data gets dumped on the SOC. Opinions oscillate wildly until all the data gets processed to fill out the picture and build consensus. Boynton says the LED counter doesn't work as well with fine particles.

"It might be tricky to sort out," he says.

I see Peter head to the kitchen and follow him. I want to know what he thinks about this new mess. I have to run to catch up, but then I slow down before I reach him to make it look like we were just headed the same way. When I finally match his step, he stops. There's a baby crying somewhere in the SOC. It's an odd sound. We both look to see where it's coming from.

"I hope that's not Heather," Peter says. He still has his sense of humor. "HQ will never be happy. There's always going to be a risk. They know that. They really have their panties in a bunch." Before he walks off, he asks, "Do you think you're getting a good story for your little book?"

"Ehhh . . . ahh . . . yes, I do think so. Thanks for asking." Peter cares.

In the TEGA office, Heather Enos looks at the data with Jerry Droege. She is not crying like a baby, and Jerry is defiant.

"It'll happen eventually," he says, "and then the sweet will never have tasted so sweet." Next step is to take an image inside the oven and then try again. Everyone will pull together for one more push.

AT MIDPOINT, THE MOOD IS MIXED.

"There's lots of data excitement," Vicky Hipkin says. It's a phrase you don't often hear.

"But it does not look like we got enough sample," she says, smiling. Nothing gets Vicky down.

"We should cheer the great rasping and the fact that we *did* get an icy sample!" she says. There's limited celebrating, however.

"We've brought some organics for everyone to sample," Vicky says and passes out a red bucket filled with chocolates.

"Bill, please go first," she says.

Bill takes the pail filled with Hershey's kisses and other treats. There's a little shovel in the pail to dig up a sample.

"Good organics," he says, making a show of eating the first piece.

The meeting starts.

Chuck Fellows, a TEGA engineer, prepared several graphs to help us understand the oven-full confusion. The graphs plot the signal strength of the diode that measures the particles. These should explain why they don't think there's a sample.

"There is a bit in the early moments, but the curve is similar to TA-4 [the first delivery that stuck]. They can't estimate the amount of material," Chuck Fellows says. He thanks the team for their hard work. He knows what an effort they have made.

The RA team reports they successfully made sixteen rasp holes. It went as planned. The SSI engineer has a series of black-and-white images of the four-by-four grid of sixteen perfectly drilled holes.

"Great work. It's just what we wanted," Bill says.

"Beautiful. Just beautiful," Vicky says. Everyone applauds. The activity was a success, but the ice delivery failed. Nevertheless, this is promising. Tomorrow we'll try again.

LONG FACES GRACE THE ENGINEERS IN THE RA ROOM.

"Now everyone will be under a tremendous amount of stress," Ashitey says. He's worried about the health of the team.

"You can see that some people aren't sleeping," he says, "but that doesn't solve problems." I think he might be talking about Bob Bonitz. I don't think he's slept for days. The black circles under Bob's eyes and his slumped shoulders don't give you the sense that he's a well-rested man. On a small mission like this, you know instantly who is upset and who is letting the stress get to them.

"I go to sleep and get up at same time. When you start to change things, you introduce errors," Ashitey says.

Images of the scoop start to stream in over the network, ending all the musing and speculation.

"Are you looking at the scoop images?" Bob Bonitz asks as he charges in.

"The sample is stuck in the back of the scoop!" he says and gives me the nod to get out. An exclamatory from Bob Bonitz, perhaps the most understated of all the engineering team, is worrisome.

PETER AND BARRY WANT EVERYONE TO MEET. NOW! WE STUFF OURSELVES into the small conference room. The big conference room is still out of commission from the storm.

"You have to leave," Peter says to me.

"What? Serious?" I say.

He frowns. I instantly regret my too-casual reaction. I forgot where I was for a moment.

"No problem," I say, trying to correct the misstep.

I can't believe I just got the boot. I collect my computer and notebook.

Then Peter turns and says something quietly to Sara Hammond. Her mouth drops open, but nothing comes out. She looks at him and then storms out.

"I have freaking DOD security clearance and I can't get in. Unbelievable," she says to me outside. I can't believe Peter just kicked Sara out too. I feel bad for her. She walks off.

I hang out with the LIDAR team. Their office is next to the small conference room, but I don't press a glass up against the wall. It would be wrong and I can't find a glass.

"Let's use the LIDAR to melt what's in the scoop," Clive Cook says.

"APAM starts in five minutes," says a voice over the SOC walkie-talkie. It's kind of garbled.

"They're calling you to the conference room. They've changed their minds and they want your opinion," Clive says.

"Quit being a rasp-hole," I tell Clive in an effort to lighten the mood. Then we sit around like Beavis and Butthead and call each other rasp-holes. We have a severe condition known as the Mission Control giggles. Another Mars Lag symptom, it's an unfortunate and sad condition.

BARRY GOLDSTEIN TALKS ON HIS CELL PHONE AS HE CHARGES OUT OF the conference room. The meeting is over.

"Let's hold off on saying anything until we can regroup and more fully understand the issue," he says into his phone.

That can't be good.

There are many closed-door meetings and the scientists ambulating betwixt them look grumpy. Finally, the group gathers for a team-wide update. I decide to press my luck and loiter nearby. This time I can stay.

"For those who don't know, there's still a sample in the scoop. It's stuck," Paul Niles says. "The new strategy is to understand the properties of this new clump and put it on top of the OM [Optical Microscope] box." This is deja vu all over again. "So now we must understand the paths," Paul says. "First, attempt to re-deliver to TEGA using the 4x4 Rasp method—we'd need to understand how to remove sample in the scoop. Second, scrape Snow White and deliver."

"Joel and I discussed that any new ice delivery should be regarded as a research project and for now we need to move on with TEGA," Bill Boynton says.

"That's the same as option two, no?" Paul asks.

"We should get away from the whole ice issue and decide the options," Bill says.

"We have to decide something now," Paul has his marching orders.

"We should deliver Rosy Red," Arvidson says. Full stop.

"I agree. We should scrape and deliver," Peter says.

"Let's not scrape or rasp. Just deliver," Bill says.

"Is this a good sample to coordinate the WCL findings?" Paul says. He wants to be sure this is a good science decision. And it'll corroborate the semi-secret perchlorate findings from the MECA team.

"It would make me happy," Bill says.

"Let's go," Ray says.

"Hold on a second," Doug Ming says. "I agree, but I want to be sure that the next sample backs up the WCL perchlorate findings. Then we have a complete science story."

"Well, we need to get this sample to satisfy HQ, and this is already in TA-0," Peter says, reminding everyone that part of the sample is already in the oven. They should consider the fact that a new sample will cross-contaminate. That could be trouble down the road when they want to publish their data. The science community may not buy a sample that has unsure provenance.

"Yes, we've got some material," Bill says, "but we're also way behind schedule. Are you considering postponing our shift to Earth time so we can finish?"

"That's another discussion," Peter says, cutting him off. There was some talk of moving the team back to Earth time in just one week. The argument for doing so is that people are exhausted and making errors—and it saves a lot of money. But let's get back to that later.

"Can we conclude this and make a decision?" Heather asks. Her request falls on deaf ears. The debate continues.

"What's the composition of scraped versus rasped material?" Barry asks. He thinks maybe they can just adjust the scraping/rasping mix to prevent sticking. Others think it more an issue of timing. They have to deliver earlier in the morning so the ice doesn't sublimate and refreeze on the scoop. Many of the scientists think this refreeze causes the stickiness of the soil. Someone even suggests that the mysterious perchlorate—the chemistry team's big find—causes sticking.

They decide the best course of action is to scrape and deliver on sol 62.

"We should go for it," Bill says.

If they just redo the plan and delete the rasp, they'll get some dry material. They can drop it in TEGA and we're back to the regularly scheduled mission. Barry says he'd like to put off any other activities until this TEGA sample is delivered.

"Maybe there should be another RA down day too," Paul Niles suggests, and everyone agrees.

Mike Hecht says he thinks Barry is wrong. They should not just focus on TEGA. He wants them to push for more MECA chemistry, too. He asks them to please add the wet chemistry, WCL, to the plan.

"No," Paul Niles says. And not because Hecht's request annoys him, but because he knows it's impossible.

"TEGA already generates a lot of DATA. There's just no room for WCL in the plan," Paul says.

"Hey, I'm 23 megs! He's 57. He's the hog," Heather Enos says.

After a brief argument, things settle. The sol 62 plan will repeat the sol 60 plan.

"Why not? Since we're so close. . . ." Bill says. The science team agrees. The RA team rejiggers their commands to reduce rasping to 30 seconds and shaves 20 minutes off the delivery time. That should get the whole process done while the sun is lower in the sky. If they can finish while it's a bit cooler on Mars, there's less chance the sample will stick.

After the meeting, Ashitey says he's surprised the science team couldn't predict the sticking.

"They've been asking us! How should *we* know?" he asks.

I tell him Nilton thought there would be sticking. His lab did a series of experiments before the mission.

"Hmm," he says. Unfortunately, Nilton is not here to say "I told you so." It's never that satisfying anyway.

More important than how Ashitey, Nilton, or I feel is what's happening at HQ. Are they going to approve the plan?

"It's Sunday," Peter says. "They're not in the office."

I guess NASA only intervenes Monday through Friday between the hours of nine and five Eastern Standard Time. Weekends you're on your own.

CHAPTER TWENTY-THREE
ICE DELIVERY, TAKE TWO

SOL 62

IT'S 7:05 A.M. LOCAL TUCSON TIME; 1:56 P.M. ON MARS. WE WAIT FOR kickoff to begin. There's one empty chair at the conference table. Where is the RA downlink engineer? After a moment of thumb-twiddling, we start.

"The RA team has a down day," Vicky says. "So let's get started."

"But they don't get the day off," Kornfeld adds quickly; as if we might think they're slacking.

The RA team needs time to regroup. They had another all-night marathon making sure the second attempt would be perfect. They are several days behind in evaluating their own data and need time to catch up. So today we'll wait for results and put a simplified plan into space, sans RA team. Today's core plan is remote sensing.

Someone on the TEGA team decided on "Shoes of Fortune" for the sample name. What does that mean? They should consider hiring a copywriter to help with nomenclature. It might do wonders for the PR effort.

"We're still all on tenterhooks," Vicky says, although there's noticeably less electricity than the last delivery. She says we don't have a decision tree laying out our next-best options if this doesn't work.

There would need to be some major rethinking, which no one has the energy to consider at this moment.

"We're going to stay positive," Vicky says. The first MRO data pass won't come for a little while. One amazing thing to note is that when everything is optimized and the deep space network works, the data comes down from space at up to 5 mbps. It's broadband from Mars.

Peter Taylor is a researcher from the University of Toronto. He's today's weatherman. He holds up a smiley yellow sun.

"It's clear and sunny on Mars," he says. Finally, a real weather report. That's not the only point of interest from the MET team. It's fall on Mars, and the sun starts to go down earlier and earlier each sol. With the first dark nights, after the autumnal solstice, they were able to take images of the LIDAR's laser beam from the SSI camera.

The green laser flickers and illuminates little bits of dust and debris in the Martian atmosphere. You can make out the hints of a wispy passing cloud. We try to take it in. Laser. Mars. Clouds. Wow.

AN HOUR LATER, DATA STARTS TO COME IN FROM MARS.
"TEGA safed," Joe Stehly says. It's the same story all over again. They start anew the process of looking for what went wrong this time.

"Your story is ruined, huh?" A young engineer asks, patting me on the back for consolation. This is getting way too meta.

Barry Goldstein and Richard Kornfeld walk through downlink.

"I have a call with Peter and Charles Elachi (the head of JPL) and I am resolute that we're going to move on. RA will have a down day and TEGA is out. We do one more try and then we will move on," Barry says.

"YOU'VE PROBABLY HEARD THERE IS NOT A LARGE SAMPLE. LET'S get reports and hear from Peter Smith about where to go from here," Vicky Hipkin says to start midpoint. Peter stands and takes the microphone.

"We've had a little time to think about the second non-delivery of ice," Peter says. He's getting ahead of himself. He pauses, collects his thoughts for a moment, and starts his talk over.

"We tried to develop a step-by-step process and sample the stratigraphy [layers of dirt] and leave the [delivery of] ice until last. But the tiger team [TEGA wiring experts] report forced us to treat each TA cell as our last. So we diverted. Many on the science team thought soil above the ice was best. But HQ dictated we get into the ice. And since July 1st, we've done it. It's become a science experiment. Now we need to get dry soil into TEGA. That's what the science team wants. And I've emailed Ed Weiler [head of the science program at NASA] and I hope to speak to him so we can be released from this requirement. We will then sample the dry soil. Probably there won't be ice, but we can *still* get results. The public wants to know our results. We need to complete our WCL story and present it to the world," Peter says. Then he pauses for moment.

"Are there any dissenting opinions?" he asks. There are none. Instead, there is thunderous applause.

"What if NASA disagrees? Is there a fallback position?" Peter Taylor from the MET team asks.

Ramon de Paula is in the room. I'm not sure Taylor realizes this. Maybe he does. I don't think anyone cares or feels like he adds any value to the situation anyway. Ramon raises his eyebrows but says nothing.

"If they're not on board, we could all take a few days vacation," Mark Lemmon says with a smile. It would be a good old-fashioned sick-out. Stick it to the man.

Peter Smith stands with an arm on his hip. He looks stoic. Peter takes over. This is his mission once again.

"I guess it becomes a game of hardball at that point. I think we're all very certain of the path we need to take. I think I was very direct about it in my email," Peter says.

"I don't think he minced words," Bill Boynton says with a side-long grin.

CHAPTER TWENTY-FOUR

MARS MAN FOREVER

SOL 63

PETER'S ASSISTANT, FRANKIE KOLB, ASKS EVERYONE TO PLEASE head outside.

"We're taking a team photo," she says. After yesterday's rousing speech, it seems like a great moment to document the mission for posterity!

I grab my camera and follow the team.

"You can't just take a picture," Rolfe Bode says. "Aren't you going to join us?" Then my inner imp takes over. I imagine the photo hanging on the wall in the Explorer's Club fifty years from now . . . and I'm in it. My heart races. Why not?

I make an executive decision: I'm getting in that photo. As long as I can stay away from Peter and especially Sara, everything should be okay. I take deep breaths to try to slow my heartbeat. The thumping noise is a dead giveaway. Getting kicked out of the photo, a public show of no-confidence, scares the bejesus out of me. I don't think my ego could handle it, and then I'd just sit on the curb crying while everyone else smiled and said "Mars!"

I kneel casually in the third row far from Peter and keep my head down. The photographer looks through his lens and then he begins the slow process of repositioning those he deems out of place. I mentally plead with him to leave me alone. I'm very happy where I am. And, of course, he doesn't listen. I can feel him looking at the top of my head.

"Hey, you!" the photographer says.

Oh, please don't be talking to me.

"I can't see you, I need you to move up," he says. I ignore him to no avail. He comes over, and, with two firm hands on my shoulders, I'm politely repositioned. I keep telling myself not to look at Peter. Not to look at Peter. Then, of course, my rubberneck car-crash brain takes over. I look over at Peter. He catches my eye, but then looks away. Maybe he didn't see me. Maybe he did.

1 . . . 2 . . . 3 . . . click. Now I'm a part of the Phoenix team . . . forever. Only an act of Photoshop can change this fact.

BACK IN THE COOL WINDOWLESS SOC, WE GATHER FOR THE END-OF-sol science meeting.

"It's been a long summer," Peter says, and waits for everyone to quiet.

"We've been through almost two Mars cycles. But we cannot continue to operate like this. There are students, teachers, children, and we have to return to Earth time. I pushed back on Barry and we've agreed to go to August 11 as a date to return to Earth time because we're very close to mission success. But we will do a transition and we will pay a penalty. Barry will speak for JPL. He holds the contracts with the co-investigators. If you have concerns, now is the time to share them." Peter then takes a seat while this news sinks in.

"What about remote operations?" Carol Stoker asks. Also known as distributed ops, this phase of the mission happens when the SOC is liquidated and the mission conducted via phone calls and Internet at a highly reduced pace.

"We will go remote on the 24th of August. We want to feel comfortable with Earth time ops first," a senior engineer on the phone at JPL says.

"That coincides with sol 89," Peter says. "The end of the primary mission."

NASA and Peter scheduled the mission for ninety days. That was a prediction for how long the lander would likely last and, therefore, how long NASA would fund operations at Mission Control. The transition to Earth time and then remote operations will happen in a few steps to keep everything running as smoothly as possible. Of course, everyone hopes it will last much longer, but NASA hasn't committed any additional funds. We don't even know how they'll react when Peter tells them he's making one last effort to get ice and then we are moving on no matter what.

"Our funding for the extended mission will go further when we move to five-day weeks. Obviously we can't keep working seven days. We'll need fewer redundant staff then, too," the engineer on the phone says.

"Are we supposed to charge for the weekends?" Bill asks jokingly.

"Yes!" Peter says. "Please send the bills to Barry."

"I will file them appropriately," Barry says with a chuckle.

"What about remote login and the applications?" Carol Stoker asks.

There are already problems with the secure accounts required to log in to the remote operations system, and anyone who has tried to participate in the mission while away complains about the software. The lag time for monitoring what's happening in the SOC can feel worse than the delay from Mars. Carol's question is, as usual, leading.

"We're working on it," the voice on the phone says. "We know there are some bandwidth issues."

"How do we deal with the nine time zones?" Stubbe, a German scientist, asks.

Mike Mellon shakes his head. He thinks these concerns are important but skirt the real issue.

"How are we gonna get the science done?" he asks.

"We took a big hit of twenty-five days and the right thing to do is push back," Barry says, trying to convince Mike the extra week of Mars time is a gift from NASA and will be enough to make up for the twenty-five days lost looking for ice. Mike isn't convinced. An extra week of working on Mars time doesn't feel like they're "pushing back" hard enough. Mike feels like the mission will get shortchanged by a bureaucratic decision.

There's a lot to consider in this move back to Earth time. First, there's a budgetary concern; NASA hasn't committed any additional funds. Peter and Barry still need to make the case for an extended mission. The support staff is exhausted; everyone is exhausted. Working on Mars time is really taking its toll. Most importantly, the team members have commitments back home. Most of the scientists teach or have labs to run, or both. The semester is going to start in September and they will have to leave. If they hope to keep the mission going after the team returns to their home institutions, they need time to transition to remote operations. That won't be easy and can't happen all at once. If they don't do it slowly, it could cripple—rather than merely slow—the mission. It's not just a matter of people calling in on the phone and continuing the mission. Joel and his team need to restructure the entire planning cycle and then the team needs to practice this new structure.

These Earth time debates might come as a shock but they percolated quietly in back rooms for the last couple weeks. There were just more important concerns. Now we hear it public for the first time and it just feels wrong. Sure there are lots of practical reasons, but this is Mars.

"I appreciate the week, but what then? What about the science?" Mike asks again.

He's polite but unhappy.

"Well, we have a boundary condition; people have to go home. It will have to happen," Barry says.

This doesn't satisfy Mike. There is another hour of back and forth. People have a lot of questions about how any work will get done on Earth time, especially once different time zones are factored in. It's messy. The scientists are angry. One sol will carry out slowly over two or three Earth days. That's going to quash any ambitious plans the scientists had for exploring the trenches and completing all the TEGA and MECA experiments that were originally planned.

Perhaps we would not have done all the experiments anyway. But spending all that time on a futile directive—from bureaucrats that never even bothered to come visit the mission—prevented us from pursuing a more logical, science-driven approach. That makes this return-to-Earth-time move feel a little more tragic, a missed opportunity for further discovery.

Today is NASA's 50th anniversary. No one celebrates.

CHAPTER TWENTY-FIVE

THE THIRD TIME

SOL 64

PETER'S OFFICE DOOR IS OPEN AND HE'S SHOWING SOMEONE OUT, a journalist. But it's not a friendly good-bye.

"This isn't *Capricorn 1*!" Peter says to this reporter who suggests they should really have accomplished more in their time on Mars. *Capricorn 1* is a fictional film about NASA faking a manned Mars mission and then murdering a bunch of people in a cover-up. O. J. Simpson is in it. Download it today.

"We're really up there doing it!" Peter says and walks off.

He just kicked out a reporter. I think he's a bit tired of getting kicked around by NASA and those darn media folks who think we're sitting around twiddling our thumbs. Now he's kicking back.

KICKOFF TODAY IS VERY MATTER-OF-FACT. PEOPLE CAN'T BE FUSSED FOR complete sentences or long explanations. Doug Ming goes through core plan: Rosey Red, blind delivery, and WCL. Then TEGA bake. Also, Trench A is renamed Cupboard. Why? Because you find great things in the Cupboard. Doug quickly lists the forward dependencies for TEGA. There's no discussion.

I don't know if this is a sign that everyone knows the routine so well by now that they've melded into a single organism, or that they're all just frustrated and tired of one another and can't be bothered to react. It is the third and final attempt, a repeat of the last plan with all the kinks ironed out (hopefully). No matter what happens, this is the last go. We'll tell NASA that we tried three times and that the ice on Mars keeps sticking to the scoop and there is nothing we can do about it. If NASA wants us to proceed, we can acquire and deliver more general samples in a systematic way. We're moving ahead regardless.

The spacecraft team reports that everything is nominal, their word for normal. Doug tells the MECA team to scrub their plan better, meaning he thinks it needs to be cleaned up. The APIDs don't look right to him.

"These APIDs are not in shape and we don't want to lose any data," Doug says. But before he can describe the problem with the data prioritization, he's interrupted.

"We have an oven-full signal!" Matt Robinson says, bounding out of the RA office. He always seems to be around when there's good news. He jumps and smiles like a madman. There's a kind of shock. He seems like the only one who really cares. He's the only one who has seen the data downlink. Maybe no one believes him.

Line Drube and Morten Madsen poke their heads outside of the RAC office. They want to know what's happening.

"No one is jumping for joy," I say.

"Well, they are all reserved," Line says. "They've been burned twice." Mads Ellehøj comes into downlink with a coffee.

"How's it hanging?" Mads asks. We get to tell him the good news.

"Oh, that's great!" he says and leaves to get Bill's opinion. He comes back a moment later.

"Yeah, it's full," Mads says.

Peter comes out of his office; gets the news. He doesn't really react either.

What's happening?

The RA guys go back to their office to work on the next delivery. No one celebrates.

"Why should we celebrate?" Richard Volpe asks. "We just repeated what we did a month ago, except now with scraping and ice." Those guys are hard to please. Even so, this is fantastic.

Heather Enos comes into the RA office to thank Ashitey for the all-night effort. She looks happy. That's two people. He goes to give her a handshake; she gives him a huge hug.

Ashitey sometimes errs on the very formal. The other night when they'd missed another meal while planning the delivery, the TEGA team bought pizza. Ashitey insisted on paying for the slice he ate. He's not allowed to receive gifts. It's part of his government employee contract with JPL, and he certainly would not want any appearance of impropriety.

"I guess the third time is a charm," Ashitey says to Heather.

HEATHER ENOS DISAPPEARS FOR A MOMENT. THEN SHE COMES BACK with cups.

"You know I pulled out two gray hairs yesterday," Heather says.

"I thought you were giving a toast," Peter says. She thanks everyone for the late nights and repeated efforts. She starts to cry before she can finish.

"We don't want to forget those people in the back room [shift II engineers] who do something to get the commands to the spacecraft," Bill says, finishing Heather's toast. We drink to discovering ice on Mars and getting back to the mission.

Heather asks what the sample is called.

"Wicked Witch," Doug says.

"Oh," Heather says. A little surprised. And not all that thrilled. Bill named it while she was away yesterday.

"I figured we were going to put it in the oven and cook it, so it seemed good," Bill says sheepishly.

"Geez, you take one day off. . . ," Heather says.

"You could have called it gingerbread cookies," Peter says, making the point that cookies are more likely to go in an oven than witches. But a sample by any name would be just as sweet.

AT THE MIDPOINT, BILL BOYNTON TAKES A SEAT AT THE CONFERENCE table.

"This is Chuck's job," he says. Heather stands behind Bill with her arms on his shoulders. He's going to present the engineering data today.

"It's a noisy signal and we're at the limit of our detection range, but we see a peak in the calorimetry. We see melting up to 18 degrees. The temperature probe is outside and so this is indicative of a little ice and a porous sample!"

Heather's eyes well up with tears as Bill speaks. His voice is full of contentment and pride. Not only did they get the sample, but it actually has ice in it! All assumptions are off, we've got ice and detected it beyond a doubt. Any number of things, most notably sublimation, could have kept us from measuring the signal. But it's there, clear as day, in the chart Bill shows: melting at precisely 32 degrees Fahrenheit. This is how they know it is ice.

"When we came here, we suspected it was water and now our mass spec [spectrometer] shows we have indeed been sitting on ice for 63 days," Bill says.

"We reveal this tomorrow at 11:00 a.m. in a press conference. So, please keep it quiet, so Bill can have his moment in the sun," Peter says. He earned it.

CHAPTER TWENTY-SIX
PRESS CONFERENCE

SOL 65

HAVE A HANGOVER. SO DOES EVERYONE ELSE. WE ALL TOOK A NIGHT to celebrate and got carried away—not without just cause. Last night the TEGA team and the engineers from the atomic force microscope hosted separate champagne parties. The MET drank wine with visitors from the Canadian Space Agency. Doug Ming watched the NOVA documentary about the mission. The Danes barbecued steaks and drank beer. Walter tried to teach them all how to ride a unicycle. I celebrated with a *Washington Post* journalist named Joe Bargmann, who was here to do a story on Peter and the famed Virgin Galactic engineer, Burt Rutan. We drank too many martinis and celebrated Martian accomplishments. He gave me some advice on telling a good story. I hope it pays off.

LATER THAT MORNING, SARA HAMMOND INTRODUCES THE TEAM AT THE press conference. She's wearing a lovely business suit and it's all very formal. A little too formal. Michael Meyer, the Chief Mars Scientist at

NASA, flew to Tucson for this breaking-news press conference. He sits on the panel with Peter, Bill, Vicky, and Mark. Meyer goes first.

"NASA has agreed to extend the mission to the end of the fiscal year," Meyer says dryly. "That's about sol 124." This is great news but not the way Meyer says it. It's a little troubling that NASA's Chief Mars Scientist sounds like an accountant. You'd think he'd be more excited to announce the big ice discovery and restore a bit of glory and prestige to the Red Planet and NASA. Maybe he has jet-lag. Still, great news.

"The minimum mission success has already been met and full mission success is not far," Meyer says in his monotone deadpan voice. It's probably my hangover talking, but he doesn't even seem prepared or excited. He just kind of flounders. I mean, why would the chief scientist at NASA talk about funding and fiscal years when we're here to talk about how our concept of another planet just changed forever? If we've learned anything from the *Ghostbusters*, it's that you shouldn't cross the beams, or if you're not a fan of *Ghostbusters*, don't mix your message. The science guy shouldn't talk about accounting, he should be talking about how awesome Peter and the team are for sticking with the sticky soil. I'm not sure why he seems so uncomfortable up there. This isn't a deposition. We're celebrating a great Mars finding. Can't they find someone that looks happy about space? We'd be better off if they put one of the interns who sit on the buckets up there. At least they're excited to be here.

Peter thanks Meyer for extending the mission.

"We've had the good fortune to land on ice," he says assuredly to the small crowd of scientists and media who showed up, and, of course, the TV audience at home. After a brief statement, he passes the conch to Boynton.

"We're really pleased to taste ice and it tastes fine," Bill says dramatically. He makes the most of this big moment. "There's only a little ice because of sticking. We tried twice and then it stuck. We didn't expect that. We decided to call the sample Wicked Witch."

Then Boynton pull out a tall and squished witch's hat he'd been hiding under the table. He talks about Hansel and Gretel and the Wizard of Oz.

"I'm melting. I'm melting," he says, quoting the dying words of the Wicked Witch. We all laugh.

Peter hates when the science team does this sort of thing. The whole mad scientist and geek thing rankles him. But it's far better than Meyer's stiff demeanor. At least Bill has verve, a point of view. It's really time for Peter to embrace a bit of that joyous nerdiness.

Vicky Hipkin shows the LIDAR movie of beams at night. She's got a great TV presence. She's exciting and clearly loves what she does. Even the bloggers agree. There are loads of comments about the "red-headed Scottish girl" lighting up the space blogs just moments after she finishes. She even gives a weather report and makes it sound fun. Summer average −3 max and −79 min.

Hipkin passes the puck to Mark Lemmon. He shows the mission's masterpiece, a great mosaic of the entire landing site. The image known jokingly around the SOC as the Peter Pan. But he doesn't call it that. Happily Ever After Pan is now the official name. I think even Sister Wendy, the cloistered art critic extraordinaire, would proclaim this photo a masterpiece. It's a triumph of human spirit created by stitching together five hundred impossible images taken of another world. We take a moment to drink it in.

Then Sara opens up the floor for questions. First, there's the standard fare. Dave Perlman of the *San Francisco Chronicle* asks if it was difficult to get the scoop of ice and if this contribution is the climax of the Phoenix mission.

"There's more to come," Bill says. Another reporter asks what does finding water mean.

"Now we can move on and look for life," Mike Meyer says.

They take one last question on the phone. It's from an industry veteran, Craig Covault from *Aviation Week*. He's been covering space for 'Av Week' for decades. This is an important trade magazine for NASA, so it's a professional courtesy to give them the final word.

"Where are you hiding the MECA team?" Covault asks.

The team laughs, thinking it's a joke. But I don't get it. And that's not the profound/insightful question anyone was anticipating. There are no MECA representatives because it's a press conference about TEGA. Peter laughs and says he's hiding them under the table. And then that's it. And our profound moment ends on a flat note. There's polite applause and everyone goes back to work. There's still a lot more to discover.

PART III

IT'S DRY FREEZE

DATE: AUGUST 03, 2008

AFTER THE PRESS CONFERENCE, THE MISSION TAKES A collective sigh of relief. We move forward free of NASA minders or bureaucratic directives. With a month left in the primary mission, the mission hits a sweet spot. We already discovered ice, maybe liquid water, imaged the landing site, made huge gains in Martian chemistry, and got a mission extension.

Peter takes a few days off. He deserves it. It's been more than two months of grueling "24.40/7" work. What could go wrong if our fearless leader takes a few days to recharge his batteries? While he's away, his number two, Barry Goldstein, seldom seen in the SOC but omnipresent on the conference phone, helms the ship.

"Barry, take the COM," I imagine he said as he shoved his three-iron into his golf bag. "If I'm not back in two days, wait longer." And then he rode off into the desert. Peter's vacation is not a romp on the beach. That's not his style. This is more of a working holiday. Peter must play golf. But it's not an ordinary match. He must defend the honor of the mission, squaring off against a rival in a modern jousting ritual . . . golf. Peter left Mission Control to play a round with a rival space

243

legend, astro-engineer extraordinaire Burt Rutan, builder of Richard Branson's Space Ship One and defender of the civilian spaceflight program. Once challenged, Peter had no choice but to go to Palm Springs and show this Virgin Galactic plane-building upstart how space is really done.

CHAPTER TWENTY-SEVEN

TINFOIL HATS

SOL 67

"I DON'T EVEN THINK THEY'RE ON SPEAKING TERMS," AN ENGINEER says when he sees Barry Goldstein pass Ramon de Paula in the hall.

"They're supposed to support Phoenix and they're barely ever here and when they are here, they seem to know very little about what's going on," says an engineer—who doesn't want to go on record—about our NASA headquarters liaisons. He's afraid of offending HQ.

Some of the TEGA team and mission managers think a few simple conversations could have settled the ice dispute amicably and without wasting a month if only Ramon de Paula and his partner Bobby Fogel understood the issues and made some modicum of effort to mediate instead of meddle. We need to ask Barry about these things.

"I'll introduce you two," Karen McBride, from NASA headquarters, says. "He's a great guy." In the past, Barry proved tricky to track down. There was a lot on his plate the last few weeks with the NASA standoff and running a Mars mission. When I mention this to Karen, she says she'd be happy to make things a bit easier. Today, I get the formal introduction.

We take our places for kickoff. Vicky Hipkin is the science lead. Today, she's all business. There are no aphorisms or science invocations. Barry stands over her and fidgets. He keeps interrupting her flow.

"I'll let Barry explain," she finally says.

"This is the 50,000-foot view of what's happening," Barry says, relieved. Barry explains that late yesterday, there was complete outage of the deep space network. The Canberra, Australia facility—one of the three Earth-based relays—went down. No one thought the plan would get uplinked.

"So now we are about twelve hours behind schedule," he says. Canberra is back up and now he wants to see if they can rush a plan through in the next five hours. Vicky says some of the activities from yestersol's plan need to be repeated.

"We may need to re-run WCL-0. But we will attempt the first AFM (Atomic Force Microscope) scan," Vicky says.

"I feel good we can move forward with the plan," Barry says quickly.

They're going to do a TEGA high temp ramp, the first AFM, and exploratory trenching; these will be the core activities. There's a new SPI I on duty. He moves slowly.

Barry paces. Sits. Stands. Rubs his hands through his hair. And wants to get on with it.

When the SPI I pauses, he seizes the opportunity.

"Are we done?" Barry asks. Kickoff finishes, but there's more to say. There's something besides the problem with the deep space network that's bothering him. Barry asks everyone to stay in their seats. He rubs his hands together.

"There's been a flurry of activity in the press. Vis-a-vis a series of meetings at the White House not regarding the discovery of water. Please don't talk to the press. This looks silly already. Refer all inquiries to Sara and Guy. This will be straightened out. There is some communication going on outside of the proper channels," he says, looking disappointed. I wonder what this is all about.

Apparently, this morning *Aviation Week & Space Technology* published a story that Phoenix had briefed the President on an important new finding but is hiding the information from the public.

"Please. Do NOT talk to the press," Barry says again.

Vicky writes a new expedited schedule on the whiteboard. There's going to be a mad dash to deliver the plan and find out what's going on in the press. Maybe someone leaked information about the MECA perchlorate findings or Nilton's liquid water.

Craig Covault, the writer who asked the funny question at the end of the ice press conference, just wrote a story for *Aviation Week & Space Technology*. The headline reads: "White House Briefed on Potential for Mars Life." We talk about the potential for habitability all the time. There doesn't seem anything too worrisome about it. Still, Barry and Vicky are stressed.

"What did you *do*?" Heather Enos asks me, in a mock-serious tone after Barry Goldstein says no one must talk to the press.

"Sorry, I can't talk to you," she says. It's not a great day for Karen McBride to introduce me to Barry.

While the story doesn't seem like a big deal, it explodes all over the Internet before you can say "series of tubes." The re-reported stories make it worse, declaring that "potential" means something far different than what it means at the SOC. Uh-oh. Many of the stories citing Covault insinuate that Phoenix found life on Mars and now we're trying to hide it! Nothing could be further from the truth.

Reading it again, you can see where things went wonky. This is primo conspiracy fodder. Technically, it doesn't mean anything that someone who works at NASA HQ could have told their buddy who works in the President's science advisory that it's possible there could be life on Mars and therefore that makes the headline true. We all know there's *potential* for life on Mars. That's why we're here. But that's not how the non-space press and tin-hatters read it. The first line of the story says, "The White House has been alerted by NASA about plans to make an announcement soon on major new Phoenix Lander discoveries concerning the 'potential for life' on Mars, scientists tell *Aviation Week & Space Technology*." That probably isn't true, but it's coming from a very important trade paper and it's already attracted a lot of readers.

A commenter on Slashdot.com, the "news for nerds" site, imagines the moment when the Phoenix scientists tell George W. Bush about the findings.

"Oh man, imagine briefing unintelligent life on the discovery of intelligent life," the writer comments. We're about to watch the birth of a conspiracy theory—from the inside.

When the world actually learns that Phoenix is not covering up an alien discovery, there will be a lot of sad bloggers, reporters, and conspiracy theorists.

If you read the article, it's pretty tepid. It says: "sources say the new data do not indicate the discovery of existing or past life on Mars. Rather the data relates to habitability—the 'potential' for Mars to support life—at the Phoenix arctic landing site, sources say." Somehow Craig Covault heard about some exciting new MECA findings. What he heard about was probably perchlorate. Only he doesn't know it's perchlorate, just something. They haven't said anything publicly about it because the team isn't even 100% sure it's perchlorate. Still, Covault smelled a good story and figured that it would be announced at the big ice press conference. Instead, there was nothing. The press conference focused on the TEGA results and ice; now he thinks there's a conspiracy afoot.

CHAPTER TWENTY-EIGHT
OY COVAULT

SOL 68

"**W**HAT SHOULD I DO IF PEOPLE ASK ME QUESTIONS?" WALTER Goetz asks me. He is concerned.

"You're supposed to refer them to Sara Hammond. But, not if *I* ask you. Then it's okay," I say. He thanks me.

There's no official word from NASA on the conspiracy. It's not really a conspiracy, but try telling that to the Internet. Phoenix is a top news story on countless fringe and mainstream sites around the globe. We can only imagine the chaos at HQ. They sit around their big mahogany table scratching their heads, cursing the Internet for spreading information with such viral-like efficiency.

"We need a plan!" says shadowy bureaucrat #1 to his subordinates.

"We'll need a few days to come up with something, sir," shadowy bureaucrat #2 says.

"Okay, get me something by early next week," shadowy bureaucrat #1 says.

"We are on it!" #3 says. "Now shall we do Italian or sushi for lunch?"

The only official word comes in the form of a couple of tweets. "Heard about the recent news reports implying I may have found Martian life. Those reports are incorrect." This is the first comment posted on the matter. What, are we in the seventh grade? The tweet comes from the Twitter account of Phoenix M. Lander, as channeled by Veronica McGregor, a tech-savvy JPL press official. For all their failings in press relations, Phoenix does use Twitter to keep younger tech-savvy fans engaged. Twitter does have some limitations, dealing with a growing NASA conspiracy chiefly among them.

Before shift I starts, several Mars conspiracy stories start to hit the social news circuit. They seem to grow in number with every refresh of the browser window. Not all of them fan the flames. Some offer interesting analyses of the situation. Keith Cowing at nasawatch.com comments: "Why is PAO [Public Affairs Office] letting a robot take the flack on this? Oh yes, the Av Week article did not claim that Martian life had been found. I guess robots do not read all that well. There is a vast difference between 'potential for' life and 'discovery of' life. *Aviation Week* is very clear on what it is reporting. Why can't NASA PAO be equally as precise in what it is denying?"

WITH ALL MY BLOG AND CONSPIRACY THEORY READING, I'M LATE FOR kickoff. I arrive just in time for Palle's weather report.

"Clear skies with some dust activity a few hundred km away," he says.

"Wow, some activity!" Vicky Hipkin says.

"I got tired of saying the same thing," Palle says.

Bill Boynton looks disturbed. He's emerged from his office and finds Barry Goldstein.

"No chlorine or oxygen," Bill says to Barry, throwing his hands in the air. "You write the press release."

There's no official response, because the team's plan for its content just fell apart. It was supposed to be simple: confirm that TEGA also sees the chemical perchlorate and then do a press conference about how amazing it is that they discovered perchlorate on Mars. The team wanted to announce that it's all a silly mixup and in fact, what they thought was a coverup of life was merely confusion around perchlorate. Bill's team quickly analyzed the data and hoped for the best. Unfortunately, science and the news cycle operate on different schedules. The analysis of the new TEGA data showed neither chlorine nor oxygen, the two ingredients you need to make perchlorate. So instead of a tidy, myth-debunking story, we have a conspiracy and another fascinating Mars mystery. Why does one sample have the new substance and another doesn't? The collective blood pressure of our serene scientists rises. There's a lot of understanding that needs to happen on a short timeline.

The chemical signature Bill and his engineers looked for—a little peak at notch 35 on their graph—isn't there.

"We looked for calibration issues," Bill says, anything to explain. "That only confirmed the calibration was perfect and the test looks great.

"If it was there, we should have seen it," he continues. It might still be there. They just don't see it . . . yet.

"Maybe ice messes up the signal, but we don't see chlorine," he says. That elicits an uncomfortable chuckle. That would be fitting. The ice in the sample they rushed to acquire could be blocking the perchlorate measurements in TEGA. I want to tell them to turn their frowns upside down because it's a great story. Everyone loves a mystery.

"I guess I better go find Mike [Hecht] and tell him we can't support his findings," Bill says.

AT MIDPOINT, DAVE HAMARA GIVES THE FULL REPORT.

"TEGA did a 1005-degree Celsius bake. The instrument is happy and healthy. It's ready to go," Dave says. From an engineering perspective, everything is great. The science is murkier. Bill Boynton would like to clarify what these results mean. He tells the scientists and engineers the abbreviated version of events.

"On sol 25 we saw an oxygen release. That suggested perchlorate, but could have been other things as well. Then WCL detected chlorine on the selective ion sensor," Bill explains.

That led the team to think there was perchlorate—an unexpected finding.

"If it was perchlorate, we should have seen chlorine in this [heating] ramp," he says.

But that's not the whole story. There is one very good reason that the results of the samples are different: Baby Bear is from the surface, and Wicked Witch is from the ice layer.

"We didn't look for it [chlorine] last time [on sol 25] because we didn't think it was there. But this time we didn't see the O_2 release either; so maybe it's because it's a different sample," he says. Perhaps it's the icy sample that affects the reading. It's a complicated picture. "Complicated pictures are sometimes very beautiful," Bill says, reminding the group that a tidy press event is not their main goal as scientists.

Boynton shows a graph from the TEGA experiment of the surface sample, Baby Bear. There's a clear peak at atomic weight 32. That's the signature for a release of oxygen. It has an atomic mass of 16, but likes to hang out in pairs. So it appears on his graph at notch 32. (Please refer to your old chemistry book if this stuff about atomic masses and stuff gets you excited.)

Then for comparison, Bill shows a graph of the current sample— Wicked Witch. There is no peak. That means there's no oxygen signal at the temperature range where perchlorate would decompose. But it's not really fair to compare the composition of samples taken from two depths. Different depths means different compositions. That's exactly why the team needs samples at different depths. They want a variety to give a complete profile of the dirt.

"If we want to compare apples to apples, we need to do another surface sample and look for chlorine," Bill says. He proposes that the next sample should come from the Rosy Red trench.

A vigorous chemistry debate ensues. By the time it's over, there are hundreds of stories in cyberspace about Phoenix finding life on Mars.

CHAPTER TWENTY-NINE
FULL RELEASE

SOL 69

THE SOC IS FULL LONG BEFORE KICKOFF. MIKE HECHT WALKS through downlink at a brisk pace, then through uplink, and back around again.

"Anyone seen Bill?" Mike asks. Hecht took it upon himself to craft his own perchlorate press release, and he wants Bill to look it over. The controversy keeps growing. The mainstream press loved the story and thousands of news outlets around the globe picked it up. There's no choice but to call a press conference. Bill isn't in downlink. Hecht hurries off to find him.

The public outreach team scrambles to prepare the press conference, quell the fury at NASA and learn everything they can about perchlorate. Right now, they're feeling shaky and want a briefing from the science team before the press conference. They already know the press is going to be confused. How? Because we're all confused.

"We will get a million requests when the information goes public, and it doesn't help that when you look up perchlorate on Google it says it's a ground water contaminant," Carla Bitter says. "We're used to people asking 'Is this even real?' for even the most basic images."

"Please provide us with a simple explanation," she begs the science team. "Otherwise, we're going to get gobsmacked." Sara Hammond comes charging into downlink.

"Absolutely no one is to speak about the results before the conference," she says. The press is hounding the science team, but she implores us to ignore them.

Bill helps Sam and Mike work out the perchlorate angle for a press release.

"They'll see it is toxic, and write that human exploration to Mars is impossible," Mike Hecht says, worried about the fate of the Mars program.

"MECA was originally built for understanding human space travel. People know that. How can we play dumb?" Mike asks. I'm not sure Hecht realizes that outside of a few dedicated journalists, most people know zilch about his instrument. Still, he's being cautious. Once the life-on-Mars story goes away, they'll need to replace it with something interesting.

Google also says perchlorate can be used for rocket fuel! Toxicity aside, doesn't that make human exploration even more possible? The best angles are considered in a sort of endless loop of science-tastic "Yeah, but what about. . . ." Sam and Mike worked themselves into a frenzy. They convince themselves that the errant news story puts their credibility on the line.

"Let's take a step back," Richard Quinn, a MECA chemist, finally says when the conversation turns to how the world will judge them for keeping humans off Mars. They don't need to discuss the future of the Mars program in the 21st century. Keep it simple and only talk about perchlorate.

"Just tell them it's premature to talk about it in any other context," Quinn says.

Sam isn't sure. Bill Boynton tells him not to worry.

"It's kind of fun that the press has misread a story that's not even true and then turned it on its head," Bill says.

Now we've got a real conspiracy. No matter what we say, doubts will persist. Any effort to clear up a conspiracy only fuels its growth. It's like proving you are sane at the asylum.

"Do you think the press release is okay?" Leslie Tamppari asks Bill.

"It's fine," he says. Sara Hammond paces back and forth. She goes to her office and comes back. NASA is really on her case. She says Ray already spoke with Andrea Thompson from Space.com, and Sara thinks that was a "big mistake." Peter talked to her, too. She doesn't want anyone to fuel the story. All she can do is worry until tomorrow's press conference.

"We should just be careful about quantifying everything and then tell the press the story we have," Richard Quinn says, trying to ground us back in the reality of the science.

The spin and timing efforts aren't really doing much besides giving Sara, Mike, and Sam ulcers.

Dr. Charles Elachi, the head of JPL, is in Tucson for a site visit. Leslie Tamppari excuses herself from the discussion to greet him.

"You've had an exciting weekend. Perchlorate. The President!" Elachi says and laughs.

"And we'll have a few more exciting days. We've got a press release and a teleconference tomorrow," Leslie says.

Aviation Week prints a retraction. There aren't many other retractions from news outlets that reported the story. Telling the world the White House wasn't briefed on secret findings just puts the story back in the news. And it makes it seem all the more conspiratorial.

Scientists run in and out of offices, Sara paces, notes get passed— and kickoff hasn't even started. We put the conspiracy out of our minds and try to get back to planning. Of course, to complicate matters, JPL planned to do its first remote operations test today. How could they have known Phoenix would get caught up in a conspiracy now? The press release is still coming together. No one asked me to give it a once-over, but I suspect it will say "We did not find life. We did find perchlorate. *Aviation Week* made a mistake. Thanks." We'll know tomorrow.

TODAY IS THE FIRST TEST OF REMOTE OPERATIONS. JPL NEEDS TO prepare for the extended mission. After sol 90, the science team returns to their respective university teaching positions and the engineers go back their home bases. The SOC will become an empty shell. Operating the lander will get more complicated. This next phase of the science and engineering teams working remotely is called distributed operations. The conference phone in downlink rings.

"Ray on the line," says the voice. Ray is the remote science lead today. He dials into downlink and watches the downlink on a webcam videoconference.

"Tabitha, can you come here?' Ray says into the phone loudly. Distributed ops will happen with webcams, remote logins to the SOC computers, and lots of conference calls. Ray conducts the first practice run of distributed ops. Ray calls over another student, Amy, and soon he has a small crowd of his students around the phone.

This doesn't seem like the most opportune time to start the trial. But the perchlorate kerfuffle is background noise for the engineers and scientists dealing with today's plan. The engineers want to move the team back to Earth time soon because the transition to remote operations is complicated. Job descriptions will change, the planning cycle will be reorganized, and, with everyone spread across the country, it's difficult.

"We have to combine the kickoff and midpoint meetings and we'll need to add several strategic positions to help the science lead," Joel says. "We're still ironing out the details." If there's one thing that JPL knows how to do, it's organize people. They use a matrix management system that gives them the flexibility to quickly put resources where they are needed. It's how they tackle logistical and technical problems so effectively. It's what you need, to get a lot of smart people working efficiently.

"Barry Goldstein has an announcement to make," Ray says over the conference phone.

More bad news?

"Today is a momentous day," Barry says. "It's one year since launch. And I baked a cake." He has a huge chocolate frosted sheet cake. Everyone applauds. Ray waits for the room to quiet.

"Okay," Ray says. "Let's get started." But he's interrupted again.

"Just a minute," Peter says. He's back in the SOC and, with Phoenix's honor nobly defended, it's time to get back to the mission at hand.

"Over the weekend there was a story about Phoenix hiding info and that the President was briefed. We have to stay ahead of this story, even if we don't know exactly what it is or where it's going. We're only part way through the science process. But we've been forced to announce our data because of leaks. So please don't speak about this. We want to keep the public in the loop, but we need to understand the complex story first. It's almost like we're being chased by the paparazzi and there are false stories. But we know that there is only *one* story to tell . . . and that's the truth. This is bad news as far as I'm concerned. I would have liked to have a peer-reviewed piece. But that's not an option."

They will announce their inconclusive results at an upcoming press conference.

"NASA has asked us not to mention the consequences for a manned mission," Peter adds. Sara Hammond repeats her request not to speak with anyone and to please not even say "perchlorate." We are supposed to say "the P-word" from now on.

"Isn't perchlorate what the movie *Erin Brockovich* is about?" someone from the outreach team asks. That would be quite a coincidence! Unfortunately, it's not perchlorate Julia Roberts is after, but chromium-6. Oh well.

"Let's get started now?" Ray says. Finally we're ready for kickoff. Today's plan is full of exploratory trenching in the Cupboard site. The geologists are excited. Forget the P-word for now; they can't wait to get their hands dirty with some unbridled Martian digging. In another development, the MECA team is finally ready to take an image with the Atomic Force Microscope (AFM). If it works, this will give us our first look at Mars on the nano scale. A nanometer is a billionth of a meter. Really small.

The crux of the plan is the acquisition and delivery of Rosy Red to TEGA. The idea is to get a synergistic measurement to compare with the results of MECA. In short, another surface sample, similar to the one that had perchlorate in it. Many of the scientists now hypothesize that the perchlorate is only on the surface.

"You're really quiet out there. You still there?" Ray asks the audience in downlink when he finishes. "The first day we tried this on the rover mission, I talked for a half hour before I realized no one was there." We're here, Ray.

This sol's plan is massive. Phoenix will work for thirty-six hours straight. It's a good distraction for our robot. We don't want her getting all worked up because of the conspiracy theory. Better off keeping her busy. Phoenix wakes up at 9:00 in the morning Mars time on sol 70 and finally goes to sleep at 6:00 in the evening on 71. There are no lander labor laws as of this printing. But our energy will get very low. The spacecraft team makes sure we keep a close eye on our power levels to prevent her from overdoing it and safing.

"THINGS ARE MOVING VERY QUICKLY," SARA SAYS. ANDREA THOMPSON from Space.com just filed a story with interviews from Peter and Craig Covault from Aviation Week.

The headline from Space.com reads: "NASA Scientist: Reports of Mars Life Finding Are 'Bogus'." Peter's quote emphatically denies that the team briefed White House officials, saying the report is "bogus and damaging."

How could this all have happened? The story starts sometime before the big ice press conference. Craig Covault insists that a good source told him that the team briefed the President's science advisory on a new Mars finding made by the MECA team. Covault figured the findings would dominate the press conference. He was surprised to find no MECA representative at the press conference.

"Where are you hiding the MECA team?" Covault had asked.

You may recall that awkward moment. If not, here's how it went:

The team laughed, thinking it a joke.

"We hid them under the table," Peter said. There were no MECA representatives because it was a press conference about TEGA. The team didn't understand the MECA perchlorate finding and weren't prepared to discuss. The press conference ended, and Covault thought he had a scoop: MECA intentionally kept from the media that day in order to hide a secret finding. The whole incident speaks to larger truths about information sharing and managing Phoenix, but that's another story.

Covault says the fury amounts to a misreading of his article. He still stands by his claim that the Phoenix team briefed the White House.

Peter disagrees.

"We did not brief anyone at the White House," Peter adamantly says. Now we have a classic Martian standoff.

And if I might do a bit more pop-psychologizing and freelance analysis, here's how it probably happened. Everyone at the SOC got excited about perchlorate. Who wouldn't mention some exciting new finding to a friend or colleague? *Just don't tell anyone else.* The colleagues shared the good news with one confidante and asked them not to say anything. Even Peter, when asked by a TV crew what the most exciting find was, said "I can't tell you." That piqued interest.

When the news gets to Craig Covault at *Aviation Week*, he talks to his super-secret sources and then they say something about the President. *Aviation Week* interprets "not telling" as "hiding." And then the Internet discovers the story and pretty soon we're the Paris Hilton of space.

MIKE MELLON IS IN THE KITCHEN POURING A 25-CENT COFFEE FROM the honor bar.

"It's not very good," he says. He's only in the SOC for three more days. He's finally run out of funding, and he's already worked through his vacations and every other break he's had.

"It's a sad reality. But I made my budget for this five years ago," he says. And he's stretched it as far as it can go. Now he has to pay attention to his other research projects.

"Now I feel really guilty about the fifty-dollar scone," I say.

"It was actually $52.00," he says. The $52.00 scone is my fault. I bled his last remaining funds. Mike Mellon's wife, Heather, works on the mission, too. We went on a coffee run last week. There was a long break in the action, so we left our Mars compound to avoid the 25-cent joe. I insisted we go to the "good" cafe. Heather Mellon drove and we misread the parking sign and got a parking ticket. The good scones cost $52.00. Serves us right for leaving Mars, even if Mike did forgive me for the scone.

"I spent as much of the summer as I possibly could," Mellon says, "and I can't neglect my other projects or students." Mike never imagined such an integrated role on the mission for himself. When Leslie Tamppari went on maternity leave, mission managers asked Mellon to help organize the Geology group, GSTG. He's an expert on Mars geology. Once he started, they realized they couldn't really do it without him.

"Plus, there weren't enough people to do tactical work on the mission," he says.

Mellon was perfect. He already understood the big picture, and they kept expanding his responsibilities.

"You can't say no to that," Mike says. "Honestly, I'd do anything to be a part of this and do a bit of science. I would have dusted monitors if Peter asked." Now he doesn't want to say good-bye.

"I'll miss it," he says quietly. "I don't feel like we finished. I'm grateful for the opportunity, but there's more work here. We're just getting to the good stuff!" Mike is most excited about digging. He wants to let the RA dig and see what's there. And we just started excavating. Sigh.

"I remember the first time I made a plan. Joel ripped it to sheds," he says. Mike thought he knew what he was doing. He worked hard to prepare for his strategic science lead duties.

"I tried to protest, but Joel would say 'No. It has to be like this,'" Mike says and laughs. You just swallow your pride and get back to work. But he learned. And he learned to love it.

"Your job is to spread the pain when you take a hit. You can't just cut chemistry observations because you're a geologist," he says.

"It's a tough job and scientists are argumentative. You forget how tough they are. They have their objective and they want to get their work done. Now I'm used to it. But then when I saw the new team rotate in, I saw it. When Paul Niles started on his first strategic [science lead] shift, it all came back." People want their work in the plan. And they're not afraid to argue for it.

"Soon they'll get comfortable with Paul and be more agreeable," Mike says. "The hardest part is that sometimes it ends up being just you and the strategic SPI in the wee hours of the morning. You try to make the best decisions you can. Then, there's no better feeling when you hand off the plan and Doug Ming or Mark Lemmon say wow, this works great," he says.

Mike says if I want, I can shadow him tomorrow. He'll let me follow him around to learn the strategic science lead job. I accept!

On my way to the end-of-sol science meeting, I pass a frazzled Katie Dunn.

"Please shoot me," she says. Today is her first day as shift lead; she got a promotion. JPL wants to give as many young engineers mission experience as possible. And this is a good setting for them to learn. The hope is that they will move on to leadership positions when the next flagship mission MSL launches in three years' time. Hang in there, Katie.

AT THE SCIENCE MEETING, PETER GOES OVER THE AGENDA FOR THE press conference. Some scientists object. Peter cuts them off.

"It's too late," he says. "The press conference is scheduled for tomorrow morning and we're not changing anything." Many on the team thought it unfair to give such little notice. They didn't have time to object. They are used to Peter's relatively democratic approach to running the mission. Some are taken aback by this autocratic "No."

Before the science team retires, the AP already has a new story titled "Martian Soil May Contain Detrimental Substance." So much for spin.

CHAPTER THIRTY

SHOVE THE REGOLITH BACK IN THE LANDER

SOL 70

I WAKE UP IN THE MIDDLE OF MY MARS NIGHT TO ATTEND THE morning (Tucson time) press conference. The news cycle operates on Earth's schedule. Participants from NASA headquarters in Washington, here in Tucson, and at JPL in Pasadena dial in and chat among themselves.

I don't even make it through the door of the Swamp, where the teleconference occurs. There's no "Shhh" finger from Sara Hammond. I just get the waveoff. Sara Hammond shoos me away. She follows me out of the room.

"There are some important NASA and JPL representatives here today. It would be inappropriate for you to attend. Sorry," Sara says. Just when I thought I was starting to feel like a part of this mission, they boot me. They want me. They don't want me. I head over to my squat/desk with the Canadians and other ITARds and dial into the teleconference. Peter comes out of his office a minute before it's scheduled to begin.

"Time for my press conference," he says to no one in particular. "Well, wish me luck."

I do.

THE PRESS CONFERENCE BEGINS WITH THE NASA PRESS MAVEN DWAYNE Brown telling everyone they got the story wrong.

Michael Meyer, the Chief Mars Scientist for NASA, is on the phone too.

"We're here to announce a non-announcement," he says. "In keeping with the open, communicative nature of the Phoenix mission, such as through blogs, web presence, press releases, and mid-mission press conferences, we have representatives of the science team here to answer your questions." And with an awkward monologue, Meyer begins.

"Today, we're opening a window into the project to allow the public to see our scientific process in action," Peter says. "Tradition is bypassed because of the extreme interest." He warns listeners against making conclusions while they are only half way through the the mission.

"We don't know the whole story, but we believe we see perchlorate in the WCL readings," he says. "Please be patient." Then he gives a little science lesson to help everyone understand.

"On the Earth, perchlorate is found in the Atacama Desert in Chile and is associated with nitrates that are mined for fertilizer. This desert is a hyper-arid environment that rarely sees rain and has no vascular plants. It is often used by scientists, as a matter of fact, as a Martian analog site. These compounds are quite stable in soil and water and do not destroy organic materials under normal circumstances. In fact, there are species of perchlorate-reducing microbes that live on the energy provided by this oxidant. Therefore, this is an important piece in the puzzle as we attempt to determine whether habitable conditions exist for microbes on Mars. In itself, it is neither good nor bad for life. There are other substances that are being looked at and can produce some of the signatures that we see in our MECA and TEGA instruments. . . . This will take some time since the individuals who would do this lab work are those same ones operating the flight instruments on Mars. But it's the highest priority to deliver a sample to TEGA that will confirm the results." Peter is firm in his statement. Maybe he's been too passive, and now they need to reclaim the story.

Mike Hecht goes next. He's less authoritarian.

"I want to categorically deny that we had anything whatsoever to do with the recent trade of Manny Ramirez from the Boston Red Sox to the Los Angeles Dodgers. That's all I have to say on that subject," he says.

Hecht then segues into an animated assessment of just how Phoenix does its chemistry. He explains how the four little MECA crucibles inject a solution into the dirt sample and then measure the chemistry with sensitive little sensors—selective ion sensors—embedded in the equipment.

Then Bill Boynton does his dog-and-pony show. It's a slightly less rigorous version of his end-of-sol science talk. He explains where they thought they might register readings as peaks of chlorine and oxygen in their graph, but they did not see them. Now they're trying to figure out why.

And that's the end of the presentation. In summation: We found perchlorate in one sample that was taken at the surface. We didn't find perchlorate in a different sample, taken a little deeper in a trench. We're not sure why. But as soon as we know, we'll tell you. Oh, and we didn't tell the President we found life on Mars.

Dwayne Brown opens things up for questions.

Craig Covault gets to ask the first question. You'd have thought NASA might blacklist him, or at least made him wait at the end of the question line.

I lean in close to the speakerphone. Several scientists joined me in downlink to listen.

"Do you think you went too far in characterizing their first chemistry results as supportive of life?" he asks, going on the offensive. He wants to know if they misrepresented themselves.

"We reported what we knew at the time," Mike Hecht says directly.

"There was nothing incorrect about what was reported. When we saw our second soil sample, that gave us confidence that we had a perchlorate signal. At that point, we went into the process Peter described: careful confirmation, careful laboratory backup tests, waiting until it was the proper time to announce those additional results. No regrets," Mike says.

Reporters from CNN, CBS, NBC, the news wires, and a smattering of international correspondents are on the line and have questions.

Miles O'Brien, one of the science correspondents from CNN, asks how they could have a conflict with the instruments.

"It's not that we're in conflict. We just don't have a handshake," Peter says.

The two results are not mutually exclusive. They just came from different locations. They need more time to confirm that what they have makes good sense.

O'Brien also wants to know how this impacts the prospects of finding life.

"I've been to the Atacama Desert, and it's pretty darn dead," he says.

Richard Quinn from NASA Ames is with Mike and Sam. He's an expert on oxidants like perchlorate and coincidentally has spent lots of time in the Atacama Desert. He says O'Brien has it wrong.

"There were some initial reports on the scientific literature that the Atacama was very Mars-like. They couldn't find organics. They couldn't culture microbes," he says. However, the story is not that simple. "But, then, as research continued and we looked a little closer, we found organics in the nitrate deposits in association with perchlorate. When we looked closer, we found microbes. So it sort of turned around when we looked at the Atacama. Now, we know that microbes can exist quite happily in oxidizing conditions. I would say that the story possibly could turn out to be the same for Mars. We don't know yet, but we will continue our research along those lines."

O'Brien sounds disappointed with the answer. Life or no life? Which is it? A questioner from MSNBC wants to know exactly what is so interesting about perchlorate.

"Perhaps someone can just kind of explain in basic terms what's the big deal about perchlorate," he says. My first instinct is to dismiss this reporter. But then I realize he represents the voice of most people trying to follow this story. They don't know anything about Mars chemistry. He needs a story from the mission that starts: "Perchlorate is an exciting new find on Mars, here's why: it's a source of food for microorganisms; it can be used as rocket fuel and it would pose a threat to humans that travel to Mars (and when we tell you about liquid water in a few months it'll explain that too)!" If we don't explain these to him and make it easy for him to feast on the Martian delight, then we fail to tell the mission's story.

Then, of course, he'd like to know if they're going to find anything else exciting. At which point it becomes a lot harder to feel any sympathy. *Anything interesting? Seriously!?*

"I guess that's a trick question," the reporter says before anyone has a chance to answer. Okay, let's get back to the serious stuff. Keith Cowing, the editor at nasawatch.com, says he has a question for Michael Meyer.

"After a decade of the Internet as a driver of stories, why is it that you all still find yourself caught off guard by this strange thing called the Internet? Why have you not had a chance to come up with different ways to do stuff? Or is this just going to be a factor of life?" Cowing asks.

"Keith, I'm actually not sure what you mean by the question. I think everybody on the team uses the Internet," Michael Meyer says in response. Oy. I put my head in my hands. What the kids on the World Wide Web would call a facepalm.

The one certainty on a mission like this is that you'll find something uncertain. If you're not sure you'll find something uncertain, then you wouldn't go in the first place. How is it NASA cannot be prepared for that? They let the media and conspiracy theorists drive the message, and then they chase them all over town instead of being open and proactive about every step. More information would help negate more misinformation.

There are more questions about the Atacama Desert and how perchlorate forms. Richard Quinn tries to undo the confusion and with the rest of the team sorts out what this all means for the future of Mars. It's mostly informed and thoughtful speculation. The results need to be confirmed before anyone can say for sure. It's fascinating that this weird place, a desert in Chile called Atacama, might turn out to be a little Mars right here on Earth. Then the press conference is over. Dwayne thanks everyone for coming.

MIKE HECHT AND SAM KOUNAVES DISCUSS THEIR PERFORMANCES ON the way back to downlink.

"I don't take anything back," Mike says. Sam agrees.

Peter returns to his office with the program scientist from NASA, Bobby Fogel, Ramon de Paula's counterpart, at NASA headquarters. *Ah, so he's the reason I wasn't allowed in the teleconference.*

"I'm glad you're here, Bobby," Peter says, patting Bobby on the back. "I like you to see things firsthand." They disappear into Peter's office.

There's still some time before kickoff. *Now it's time to forget trifling media matters and begin my strategic science lead apprenticeship.*

Mellon sits with Paul Niles. Niles served as the strategic science lead for the better part of last week. Paul debriefs Mellon. He needs to get Mellon up to speed before handing over responsibility. This is Paul's sixth crazy day on duty and he looks pale and haggard. There are bags under his bloodshot eyes. He keeps crinkling his large forehead to try to keep his eyes focused.

Last night was difficult for Paul. It was one of those sols that reminds Mike Mellon how hard his job is. Science team members complained their science was ignored while TEGA got all the resources. They said Paul plays favorites because he's a part of the TEGA team.

"I thought there was the strategic directive to get the next sample in," Paul said defensively.

Ray, calling in remotely to monitor the situation, admonished Paul and told him to be fair. He couldn't do anything right. It was rough.

"I'm happy to get a little break," Paul says.

He asks Mike to figure out the necessary activities to do another round of exploration trenches. He was going to do it, but he has too many other loose ends to tie up and he says he's just too tired. He brings Mellon up to speed on the TEGA issues.

"TEGA requested a sol 72 early wakeup for the delivery," Paul says. "But I'm not sure we can do it because of the battery constraints."

Phoenix is coming off its monster 36-hour sol, the longest yet. After the thirty-six hours of work, there is very little charge left in the battery and there might not be enough power to make the delivery. They probably need time to charge the batteries before turning on the heaters and executing this power-hungry maneuver.

Mellon says he will consult with the battery and power specialists on the spacecraft team. They can calculate how much power the solar panels will likely pull from the sun that morning. Then they'll get back to TEGA and let them know if the spacecraft has enough power for the early morning delivery. Paul and Mellon decide there should be some sort of contingency if they run into power trouble.

Then it's on to the next set of issues. The spacecraft will generate a lot of data. Too much. They need to massage the downlink priorities and cut some activities.

"We have to de-prioritize the AFM (Atomic Force Microscope)," Mike says. He demotes the second round of scans to drop-in science, meaning it gets done if there's extra room in the plan. These would be the first to go if the plan gets overbooked. I'm trying to keep up. But there's a lot to keep track of in this new job.

"You have to know what happened the last three sols and what each group in the science teams wants to happen in the next three sols," Mike says. That's six days' worth of desires for six instruments.

THE CONFERENCE PHONE BEEPS.

"Ray is online," a disembodied voice says over the PA. The remote operations test continues and Ray starts the kickoff meeting. We're waiting to find out what happened with the TEGA high temperature ramp repeat, and we're hoping to get the first ever AFM scans.

The new plan is complicated. There won't be a lot of power after the 36-hour sol. So we'll have to be efficient.

"I want to try to use the RA for trenching and work on divot imaging," Ray says.

He even considers room in the plan for more pictures of the lander legs. That will make Nilton Renno very happy. He's currently on vacation. He took his son to LEGOLAND. Even though Nilton is away from SOC, he's waging a vigorous email campaign defending his liquid water findings.

He and Mike Hecht are at loggerheads over the issue. Nilton says he's not even sure he's going to come back for the end of the primary mission. He feels like he needs to continue to experiment and the MECA team isn't willing to let him use their lab. The only way to test his hypothesis now is to return to Michigan and do the work at his lab. Nilton promises me an update as soon as possible. Hecht doesn't want to talk about it.

At kickoff, Ray insists the TEGA bake gets in the plan, even if there are power constraints.

"We should consider some de-scopes to make it happen," Ray says.

That means removing other power activities from the plan so they can get the TEGA perchlorate results ASAP. Mike Mellon shakes his head.

Chuck Fellows, from TEGA, interrupts.

"I don't think it's possible. The spacecraft team asked us to pull the TEGA bake," Chuck says. After carefully modeling the power consumption, they're worried about drawing down the batteries. They don't feel comfortable with the margin they've left themselves. Since depleting the batteries could be catastrophic, they're not willing to take the risk. Ray yields to the spacecraft team's judgment. TEGA is cut from the plan. We'll have to wait.

MIKE MELLON FLAGS ISSUES FOR TOMORROW'S PLAN, MAKING NOTES to remind himself of the open issues and potential problems.

"Data passes . . . where will the TEGA sample come from? How will they deliver to the partially opened TA-5 doors? They'll need RAC stereo of Cupboard for further trenching," Mike talks to himself in lander code.

After midpoint, the strategic science lead's day goes from merely frustrating to maddeningly complicated. (These are technical distinctions.) Now Mike incorporates data from downlink into the hypothetical plan he is putting together. Then he adjusts the plan based on the actual data and makes an educated guess about what he thinks will come back in tomorrow's plan. Once he feels confident about his guesswork, he'll make sure all the activities are validated and make a list of which blocks the team might still need if they want to execute his plan. Will it all fit in the predicted power constraints? Hopefully. Does it seem likely something will go wrong? What's the fallback plan?

Today Mike needs to identify which is the best sample for divot images. He wants input from Ray. So we go track him down. He's remote. Let's call him. This job is part accountant, part bush tracker. We hunt down scientists and then carefully notate the responses to our queries in the plan.

Mellon notices that the location favored by the science team to take a sample doesn't have a recent DEM (digital elevation map). But the Neverland footwall does. That might be a better location. This is a simple change in the plan, and might keep the already-overtaxed RA team from total meltdown. Having a thoughtful SSL can make or break your sol.

I continue to shadow Mellon, but he moves fast. I thought I was good at stalking, but I still have a lot to learn. After a quick bathroom break, I can't find Mike for almost half an hour. Space waits for no man.

AT THE PLANNING MEETING, A CURIOUS NEW ACTIVITY GETS DEBATED.
"But what if it gets stuck in the scoop?" an engineer asks. The geologists want to turn over a rock. This makes the systems engineers nervous. This isn't like ice getting stuck in the scoop. If a rock gets stuck, it doesn't melt. It just stays there and the RA is kaput. The mission is over.

"But we've already moved a rock," someone from the geology group says, referring to the traffic accident when the RA hit the rock called Alice.

"I suppose we could call that practice," Rich Volpe from the RA says. And Joseph already started the testing process. He's probably in the PIT working on it right now.

A contingent of geologists and engineers discuss the dangers of moving a rock and what scientific benefits it will yield. The conclusion: you find interesting stuff under rocks.

As the strategic science lead, Mike Mellon is an impartial custodian of the long-term plan. He's also waited five years for an opportunity like this. He has to responsibly balance his tactical workload with his deep desire to turn over a Martian rock. He very much wants this rock-flipping activity, but he's not going to force it into the plan without proper due diligence.

"You often find salts under rocks in the Antarctic," he says. Where there are salts, interesting discoveries are made, like perchlorate. Mike Hecht asks Mellon how the salt gets under the rocks. The conversation turns to atmospheric precipitants. Before we're pulled back to the realities of the long day ahead of us, Carol asks if maybe they'd be better off putting an icy sample in MECA rather than flipping the rock.

"No," Richard Quinn says, "that would not be an appropriate sample." MECA needs dry material. He thinks ice would dilute the sample in an unpredictable way and give inaccurate readings.

"I have some happy news," Heather Enos says. "We've gone eight sols without any TEGA problems." That's something to feel pretty good about. No signs of any more TEGA-killing short circuits.

Mellon says we need to move on. Testing the rock flip is already under way with our friend Phoenix II down the hall. They can make their decision based on the results.

"YOU WANT TO SEE SOME ROCK FLIPPING?" JOSEPH CARSTEN ASKS ME from behind the barrier in the PIT.

Heck, yeah!

Using the Phoenix arm to peer beneath a rock is a very slow process; arduous even. Rolfe and Joseph are on hour seven of their test. They managed three attempts so far. They are in pretty good spirits, considering. Rolfe says he's a bit scared about something going wrong.

"I'm not worried," Joseph says. "I can guarantee success under these PIT conditions."

Which, of course, is a guarantee of nothing. I don't know why he's so confident. In the three attempts made over the last seven hours, only one was successful.

"Most of the time we're just working out how *not* to push the rock," Joseph says.

"The science team says they don't want pushing. We have to flip it," Rolfe tells me.

Otherwise they will disturb the soil, and that's not good.

Carsten painstakingly resets the fake Martian scenery, carefully combing the faux Martian regolith for his next try. He measures the distance of the rock from the trench. Then he carefully grooms the dirt. And when everything is prepared to his exacting specifications, he presses the test rock firmly in place. It looks just like the images we see on Mars. He calls out some commands and coordinates to Rolfe, who works the controller.

"Okay, ten cm," Joseph says. And then "yes, three radians."

After another two hours of testing and documenting various approaches, heights, depths, and lots more of Rolfe's home-baked cookies, I decide it's time to go. I have to return to my core activity: shadow Mike Mellon. I find him in the conference room. He's still going through the science team's requests for future activities. Making sure there's an optimal number of activities, power usage, and data volume. I sit with him while he works. Like rock flipping, it's a slow detail-oriented task. But I want to soak in as much of the job as I can. You never know when they might need a backup strategic science lead. After a fifteen-hour day, Mellon feels like he's done all he can. He'll hand off the sol 72 plan to the science lead, and tomorrow they'll work through it one more time in shift I. Then they'll send it to the shift II team for coding and then transmission to the spacecraft.

MIKE HECHT AND CAROL STOKER DISCUSS THE PRESS CONFERENCE AT Carol's desk.

"He [Craig Covault] forced NASA to respond because *Aviation Week* is an important channel for them and he's a veteran journalist and generally reliable," Mike Hecht says.

Hecht says Covault had a good relationship with NASA. He thinks the article was a sort of betrayal.

Carol asks Mike Hecht if he remembers an incident when a reporter at a cocktail party overheard some drunken evidence that NASA had new evidence for life on Mars.

"She went home and ran with the story and then NASA actually issued a press release that said 'No new evidence for life on Mars,'" Carol says incredulously.

"We should be giving the media headlines, not the other way around," Mike Hecht says. They agree that the story is complicated and needs to be managed better.

"You can't do science by press release," Carol says.

The press conference seems to satisfy NASA. If Sara's stress level is any indication, things are better. The reaction from everyone else is as expected. The tin-hat folks continue to discuss the coverup, but the mainstream media no longer thinks Phoenix is hiding aliens. So while there's still a bit of lingering resentment, the story is put to bed. I follow shortly after.

CHAPTER THIRTY-ONE

THAT'S THE PLANET I SAW ON TV

SOL 71

I HAVE VERY FEW SPACE BUCKS IN MY MARTIAN ACCOUNT AND I OWE rent. I bunk with the Danes, and it would be a shame to get them all evicted because I can't cough up the rent money. Earthly problems have no place in space. But they are a real distraction.

You may wonder how a person supports oneself on a Mars mission like this. The scientists all have university affiliations and NASA funding. They file for grants, and some get help from their governments. These same pools—or puddles—of money are not available to independent stowaways like me. My lavish lifestyle eating flavored oatmeal packs and binge-drinking 25-cent honor-bar coffees is NOT the result of a NASA earmark funded by your tax dollars.

Long before Phoenix blasted into space, when I first met Peter and knew that I was going to beg, borrow, and steal to be a part of his space dream, I started saving my pennies. But I also suggested a scheme for a semi-brilliant (I thought it was brilliant; others not so much) prime-time Mars TV show, featuring the Phoenix Lander team and, of course, Peter Smith. This show would offer the world unprecedented, 24-hour, high-definition access to the mission. It would be like this book but . . . ah . . . better. It's *Big Brother: Mission Control*. And you could have watched TV this whole time instead of forcing your brain to participate in this archaic eye-straining activity.

The conceit for the show was so awesome because instead of arguing over which vapid housemate might get kicked out, our show's dramatic hook would be live Martian discovery. It couldn't fail. Smith and his team would be big stars and kids everywhere would come to science class with a new zest for learning.

"Hey, that's the planet I saw on TV!" children would excitedly say whenever they looked into the night sky. Peter was excited. The *Discovery Channel* liked it too. Then, through a little TV magic of their own, they "improved" on the awesome brilliance of the project and replaced it with something they felt more comfortable: a two-hour documentary geared at the folks who already loved the mission. They thought, wouldn't it be amazing if we just had scientists talk about the mission against changing backdrops!? "YES!" they told themselves and high-fived each other. They thanked me and said they'd handle it from there. I could see myself out. Not that I'm bitter. (That's just my usual tone of voice.)

They did offer me a consolation prize. I could work with some sweet-natured television friends of theirs and they'd pay me a small producer's fee. That would be enough budget to stay with the team for ninety sols (and give me a unique chance to be a spectator on the making of a documentary based on a TV show that never got made). Perfect.

Unfortunately, the production company they hired is filled with mean people who just bounced a check and—unbeknownst to them— just put the whole Danish team at risk by getting them kicked out of their corporate home for non-payment. Now I have to spend the morning on the phone trading insults. And that means I'm not paying attention to the mission. So on behalf of the production company— who shall remain nameless—I apologize for the distraction.

Also, the bounced check is relevant, I swear. Why? It's a symbol, a wake-up call that the fat times are over. I can't live off the old discoveries. I have to be scrappy and lean and use the next fourteen sols to learn something new. I must push past this crushing Mars lag and get at the heart of the mission. I email Peter. We need to meet. He says no problem. Excellent.

AT KICKOFF, DOUG MING WELCOMES THE TEAM. JOE STEHLY, WHO JUST last week trained to be the downlink lead, has a new job: shift lead. Big promotions keep coming to reward some of the junior team members who worked tirelessly in the background for the last few years. This gives them a brief taste of leadership roles and real Mars experience. It's Joe's first night, and he looks a little nervous.

"There are many new people in key positions today, myself included. So please be patient," he says. I detect a pleading tone in his voice. Don't worry, Joe, there are some senior managers standing by to lend a hand.

Today's plan features a new TEGA delivery. That's what we need. But there are issues, large and small. There will have to be several "go" or "no-go" decisions that must come quickly after we get our downlink.

One big non-TEGA concern is that the LIDAR (the pew! pew! pew! laser beam that measures dust and clouds in the atmosphere) overheats. Until they figure how to keep it cool, they don't want to risk burning out the laser or possibly the whole lander. But I can't help but think about a story Jim Whiteway told. Back when they were rushing to deliver LIDAR, there was a test gone bad: the LIDAR overheated and caught fire.

"I didn't sleep for two months," Jim says. Everyone is always saying stuff like that on this mission. But Jim really didn't sleep. He worked 23-hour days and his graduate students did their best to keep up. His blood pressure was through the roof and he was a real mess. Even Jim's doctor told him that he was facing serious health issues—like he was going to kill himself—if he didn't start sleeping more. JPL suggested that the Canadian team should quit; they'd never make the deadline. Jim refused. Somehow they made it work.

Are the Canadians trying to burn down Phoenix?

"The LIDAR is fine. It just gets too warm during the day," Clive Cook, one of Jim's grad students, says.

"So far, all the LIDAR activities were in the morning or afternoon. It's these midday events that give us trouble. So, it's not unexpected that it's overheating. It has a lot of thermal protection," Clive explains. This insulation protects it from the cold. The LIDAR can't take off these coats when it gets too hot.

"Wow, the data story looks great," Doug says, looking at the PSI plan for the next sol. His comment elicits a smile from Mike Mellon. It only took him fifteen hours to make sure everything fit. And I was there to shadow him!

THE LAST SOL ON MARS TIME WILL BE SOL 76 (PLANNING SOL 77). JUST a few days from now, when we go back, things are gonna get a little weird. And you should be prepared for that. Because we're out of sync with the Earth clock, timing is tricky. As it stands, shift II on sol 76 ends early in the morning. So if we started on Earth time the following sol, shift II would end and then immediately start again.

"We have to have a little compassion," Joel says with a smile. So in the name of mercy, we're going to skip part of a sol. There's a half-day transition. Sol 76 is a Sunday. And we'll take Monday morning off, so shift II can sleep for a few hours. Then on Monday, shift I will start at 12:00 p.m. Shift II will start a few hours later.

Then Tuesday, we will get to work at 9 a.m. and the shift will proceed just like a regular Earth day. But it will not be a regular day for us. We work while Phoenix works. This is different than usual. So far, we worked while Phoenix slept so the lander would have a fresh plan on waking. Now that we're out of sync, it's called a restricted sol. On a restricted sol, the plan is sent to Mars without much knowledge of what Phoenix actually did versus what was planned. For instance, if the RA safes, we won't know it: we uplink blind to the results of the RA activity. There will be times when data starts to come down just as they are about to send up a new plan. And the new plan will be misguided or contain activities for instruments that are safed (and are un-executable). It will be tempting to make last-second adjustments. Be warned: it's dangerous.

"We'll have to exercise restraint if we find something cool," Doug says. There's a lot of concern, and some anger, about how everyone will do their jobs on Earth time. Transition is complicated. Team members are tired and weary from Mars time, but the mission will slow dramatically when we move back. We might cut our productivity in half.

"It's the humane solution," Joel explains. "We have to structure sols for earth time and deconstruct certain jobs so that they make sense." There all kinds of odd problems cropping up.

"It's not often you have to distribute the physical hardware of a mission," he says. We have to figure out which facility owns which computers. Plus, the IT team configured these systems to account for ITAR restrictions and SOC-based users. That has to change. Then there's the issue that none of the planning software was optimized for a distributed mission. The software all resides on computers at the SOC, and it requires some additional features if hundreds of users from all over the world need to access it remotely. The software was coded this way to save money. But now all the band-aid solutions to get Phoenix on the ground with very little money are proving very painful to pull apart. They have a shaky infrastructure and a great mission, but that's better than great infrastructure and no mission.

Ray Arvidson, who is telecommuting in from his office at Washington University in St. Louis, says he can't hear over the phone.

"Speak up!" Ray says.

There is a funny noise coming from Ray's end. It sounds like he's jumped into a pool.

"Ray might have just been abducted by aliens," someone says. Everyone laughs.

"They're finally taking him home," Bob Denise says. Or the remote operations infrastructure is already collapsing.

CHAPTER THIRTY-TWO

FORTY MINUTES BACK

SOL 75

IT'S SUNDAY NIGHT AND OUR LAST DAY ON MARS TIME. I HAVE MIXED feelings. Abandoning my Mars watch will relieve the brain cloud, but the renewed ability to think clearly is tinged with sadness. With an untested new workflow and so little time left, there likely won't be any more big moments of dramatic discovery. I spend my time thinking about this bittersweetness while I wait for Peter. I have lots of time to rhapsodize, as he's over an hour late.

I spend more time thinking. Mars time, the mission, how I can get my Peter Smith scoop and feel good about my work here this summer.

"Are you ready to order?" the waiter asks. Then I have dinner. And I wait. And another hour.

"Can I get you another?"

Sure. No problem. And dessert.

Peter is two hours and forty-five minutes late for dinner. I think he forgot. But it's a nice night.

I probably should have been at the SOC anyway. That's the lesson here. There's always more work to do at the SOC. When you leave, bad things happen. Besides, three hours late for dinner isn't so bad.

Peter arrives and my resentment is fleeting. He has a good excuse about mission management or something. Soon we just hang out and drink. We're joined by a colleague, Catherine Patterson, the woman who introduced us back when Phoenix was still being assembled on Earth. We talk about his golf trip, the NASA controversy; we catch up. There is even one point when Peter distinctly says he likes me. It was in the context of not wanting to say too much in front of reporters for fear of making a fool of himself.

"Reporters can take what you say out of context," he says. "Sometimes I'm just too open," he says.

I certainly haven't found that to be the case. If he thinks he's been too open already, there's probably no chance we'll ever have a big heart-to-heart. This is worrisome. But still, he did say he liked me. That's something. With just fifteen sols left in the primary mission, there's not much time for bonding. And Peter doesn't do a lot of emoting when we talk. There was that one inflected pause back on sol 10. Maybe there's still hope.

How can he be the hero of this book if he's not going to spill it? As I sip from my fourth scotch, I start to think a lot about gravity. Maybe he's distant for a good reason. He doesn't want to take credit away from the rest of the team! Peter is a quiet, calm, clock-in-the-storm leader. He picks the right people—those who know more than he does—and steps back while he lets them work. Some of the JPL engineers who are used to iron-fisted management express their unease with his relaxed leadership approach. But it seems to work. And with this evasiveness, Peter tricked us into learning about him through all the people around him. The portrait of Peter is created through negative space. It's an optical gambit from the premier Martian photographer.

THE NEXT MORNING, WE ARRIVE FOR SHIFT. IT'S NEARLY NOON LOCAL Tucson time. I have a headache. It must be the shift back to Earth time. Tom Pike is at the back door of the SOC. His security code doesn't work. He tries again. Nothing. This outer SOC door has a numbered keypad. Then you have to swipe your magnetic security badge to get inside. He looks up to think for a moment.

"Geez, what am I doing?" he asks. "I put my bank code in." He laughs. I tell him I won't publish his bank.

"Earth time is here. The end is near," Tom says with a smile.

Over the next few sols we'll shift back to Earth time. There will be more remote operations tests and the mission will endure.

Inside the SOC is something curious: sunshine and windows! Someone pulled down the blackout paper that's covered them for the last few years. The paper was there to help us live on Mars time and minimize the sun's impact on our circadian rhythms. Artificial lighting helped reduce the pain of living on the Mars clock, but now it's all sunshine at the SOC. A sign on the wall reads, "Back to Earth."

"Hooray, it's earth time!" Ray says and starts kickoff.

"Sadly, it's a restricted sol. Today we'll be pretty limited," he adds, damping any celebration. All the scientists have mixed feelings about the transition.

There's a simple plan and tired engineers. They work quietly. The SOC clears out early.

CHAPTER THIRTY-THREE

SCOOPED

SOL 77

THE FIRST TEST OF REMOTE OPERATIONS COMPLETED, RAY ARVIDSON is back. He returns to the SOC to lead the science day in the flesh. His step is lighter and it almost looks like he's smiling. He brings good news too.

"We need one more subsurface sample delivered to TEGA in order to get full mission success," Ray says. Barry Goldstein interrupts.

"Ray, hold on," he says over the conference phone, having dialed in to the meeting from Pasadena. "We *have* met full mission success by the letter of the law."

Ray shrugs. Silence.

"Ray, call me," Barry says.

"I can't, Barry. Why don't I have Peter call you?" Ray asks, giving him the brushoff. He has too much work to do. Or else he doesn't like talking to Barry.

"I think we have a disagreement at the SOC," Peter says.

"That was Peter, not me," Ray says, hinting at some behind-the-scenes tangle over whether to declare the mission a success. These upper management tensions are rarely seen by the rank and file down on the line. It's bad for morale.

"I will call you, Barry." He doesn't seem that excited. That's probably because "success," like beauty, is subjective. This declaration happens to be in the eye of the contract holder.

JPL gets a financial incentive if they declare mission success by a certain date. Although, "full success" and "letter of the law" success are hard to parse. I promise myself I'll dig through the 55,000 Phoenix requirements to figure out what the discrepancy is, just as soon as I have a free moment. Chris Shinohara says this is a problem of perspective. It's how you see the mission.

"JPL is great for project management, but sometimes they miss the point on science," he says. They are structured around engineering successes. That's at the heart of why Barry declared mission success. And why he says "to the letter of the law" when he says it's appropriate for us to make this declaration now. We achieved technical success, but we're not where the scientists want to be.

"Peter wants to be sure the scientists are happy and getting what they need. Peter is good at that. He's a great politician," he says. "But he can't and shouldn't upset the partners [JPL and Lockheed] because that's his job—to keep everyone working together."

"Barry, on the other hand, has a way of occasionally ruffling feathers to get things done," Chris says.

Chris makes a good argument. Declaring success now just seems like a cop-out. There are still five remaining TEGA cells and no sign of a short circuit destroying anything. It's time to dig in and explore Mars. There's work to do. No need to trumpet to the media that we're a success. Let's keep going! Already the days are getting shorter and the robot arm is showing some signs of wear. This isn't the time to quibble about mission success.

EVEN STRANGER THAN THE DEBATE ABOUT PHOENIX'S accomplishments is that Craig Covault, the *Aviation Week* writer and father of the great Phoenix conspiracy, is in the SOC! I'd say it's some covert operation, but Sara sits next to him in the visitor's seats. She must be able to see him. He listens intently to the debate; he takes notes; he scratches his head. There's nothing too suspicious or conspiratorial.

Throughout the day, everyone asks Sara the same question: what the hell is he doing here?

"He wants to set the record straight, to explain his story," Sara says. "I asked the team if they objected, and no one did." Sara throws up her hands saying, in effect, it's a little late for objections now.

Sara leads Covault into the small conference room. Covault has a slow, intentional gait, giving the short trip the feeling of a condemned prisoner's last steps. This isn't supposed to be a lynching. Covault asked if he could give a short talk at the end-of-sol science meeting. If you thought the Nilton Renno science smackdown was popular, this is even bigger. Covault stands at the front of the room, eyes fixed on the carpet below him. Scientists flow into the room. Then Craig lifts his head and begins. I thought he'd be so much more fiendish. Instead, he looks like a regular pleasant guy with neatly parted gray hair, jowly cheeks, and standard-issue wireframe glasses—a nice gentleman who has one great job, writing about space. I'm even a little jealous that it's his first day here and he already gets to give an end-of-sol science presentation.

"One of my colleagues told me that MECA had some interesting stuff going on and that there would be a press conference at HQ. Then at the press conference nothing happened," Covault says, beginning his explanation. He assures the team that he had three reliable sources and that he confirmed the White House science office had been notified. There are some groans.

Peter ambles in. He walks through the conference room. And when he's nearly at his seat, he pauses.

"Hi, Craig," he says. And then he sits.

"Hi, Peter," Covault says, "I was just explaining what happened."

"This should be interesting," Peter says, leaning back in his chair. It's very brave and somewhat noble for him to come set the record straight. Being a coward, I know in my heart that when I fabricate the story about aliens in chapter 42, I'm going to run like hell when the team comes looking for answers. Not Craig Covault; he's here for reconciliation and truth.

"I used the word 'notified' in my headline. It was the desk editor who changed it to 'briefed,'" he says. Clarifying this first point of misconstruing the scientific facts, Covault argues it was the word "briefed" that got the Internet buzzing.

Covault says that while that wasn't his fault, he does take responsibility for using the phrase "holding data." "I am aware that Peter took issue with that," he says contritely.

Covault says as of today there are 2,300 citations for the story. He never expected such a huge response. And a lot of the attention came from misreading what the story actually said. That's a valid point. A lot of the mushrooming happened from these misreadings.

"And that's how something small can turn into something big," he says in conclusion.

Are there any questions?

"In your business, do people understand the difference between 'not ready for publication' vs. 'holding'?" Bill Boynton asks with just a whiff of smugness in his voice.

"I understand. And I blew past that traffic light," Covault responds.

"Saying we were holding and going to the White House actually makes it a false story, Craig," Peter says.

"There was already a lot of buzz about a discovery in MECA and it came down to a wrong choice of words," Craig says, but there's not a lot of spirit in his half-hearted defense. He's not here to argue. He just wanted to explain how things happened from his perspective.

No one is happy with that answer.

"What will your next story say?" Peter asks. Boynton says it'll probably be a bigger story, but they're not ready to talk about it. Covault senses the crowd has turned.

"Thank you for letting me come," he says.

Sara Hammond leads him out of the room. There are a couple of science presentations about dust and chemical analysis.

Barry asks if they're finished. He'd like to give a briefing on Phoenix's power consumption. From his office at JPL, he tells us it's time to consider what Phoenix's last days will be like.

"Starting on sol 93, we'll start to see a decline in power," Barry says. It's a kind of morbid talk about how much time our lander has left. After about 120 sols, things will be very limited. And there's no hope after 160.

SHIFT II, FOR THE MOST PART, HAPPENS IN THE UPLINK ROOM. THIS
is where scientists' brilliant ideas become sets of ones and zeros:
digestible bits of instructions for the lander. It's a nondescript
room a few paces beyond the front desk in the SOC. There are different, otherworldly species of engineer who dominate the uplink
landscape. There are sequencing engineers, SPI IIs, and even SPI
IIIs. These engineering species rarely come to the shift I downlink
world. They'll pop in briefly for the second part of the midpoint,
but this is their natural habitat. It's a whole other engineering ecosystem. Here there is little conjecture or debate. Rules rule here.
This is where Julia Bell makes science happen on Mars. For all
its formality in operation, it might surprise you to know that the
uplink room is a far more intimate space than downlink. While the
atmosphere is more intense, there's a greater sense of camaraderie.
There's lots of gentle ribbing and feigned insults. More fights. More
jokes. Mistakes come at a higher price, and that makes it a difficult
but rewarding place to work.

"You leave for a day off, and when you return six more flight rules
were written," a shift II engineer tells me. "And you better know
them all."

When midpoint II breaks, the shift II workers get to work. They have
a series of more rigidly structured meetings, the Activity Plan Approval
Meeting (APAM), Sequence Walkthrough, Final Sequence Integration,
Sequence Validation Meeting, Command Approval Meeting (CAM),
and finally the Validation Approval Meeting (VAM). This is the sausage grinder Dara Sabahi spoke about. Then the mission manager and
shift II lead sign the plan and it's radiated out to the orbiters and then
down to the lander.

Right now we're at the Command Approval Meeting; we're CAMing.
It's pretty close to the end of the shift II day and the plan is almost
finished. For CAM, each instrument team has a series of checklists to
complete. Then they'll explain their activities to the mission manager
and shift II lead.

Ashitey works on a movie of the dig plan. It helps get you into the
mood when you watch these animations of the RA digging around
on Mars.

"Woah, who is that woman?" I ask Ashitey.

There's an animated woman standing on top of the lander. Holy cow! There are people on Mars and they are damn sexy.

"Oh, she's not on Mars," he says and laughs. Now that's a conspiracy! Okay. It's not really a woman on Mars. This animated lady is a vestigial element of the software the RA engineers use. In an effort to save money, the RA team and programmers at JPL use free, open-source animation software to get the playback features on a budget. Occasionally a software bug produces a female figure on top of the lander. Even with the lady, the animations are useful. Visualizing the scoop operating in the Mars environment lets them present their work and discuss any possible dangers with the science team. They show the video instead of listing the vector angles or any of the other variables in their matrix. Vector angles can have a narcotic effect, inducing sleep in all but the hardiest scientists and engineers.

The movie is a good lesson in storytelling. A huge annoyance for Peter is the team's near-ubiquitous reliance on needlessly complicated graphs and technical jargon. Peter knows it's all about how you share data that makes it relevant. Vectors and radians are hard for non-RA engineers to visualize; a scooping movie is easy. If you lose your audience, you lose the activity and maybe the mission.

The RA is the only instrument that could easily damage the ship or get dirt in sensitive areas; so there's an extra measure of caution for approving its activities, and that's why they have extra tools to visualize what they're up to. The movie lets everyone quickly agree on the digging and delivery path.

The plan development proceeds through each evaluation and test. The process becomes more rigorous at each meeting. They run several computer simulations of their code, putting it through the paces on simulated landers. The shift II lead and mission manager grill each sequencing engineer about how their code might risk the health and safety of the mission.

Bob Denise is mission manager. He polls the team for possible problems while he reviews each line of code in the computer-simulated mission. The Predicted Events File (PEF) provides a running commentary on the computer simulation. It's a computerized cop that tells Denise and the instrument teams when the lander violates any flight rules or undertakes any suspicious activity.

This questing puts pressure on the sequencing engineers to ferret out potential problems early on in the process. By the end of the evening, they should have code that's rigorously tested, they should understand possible constraints and potential flight rule violations, and they even guess at what they believe is the top risk of the day.

"The top risk for RA is that we faulted on 77," Ashitey says. He worries that the encoder drift problem will likely cause the RA to safe. He explains that they risk getting nothing and might waste a sol. Every engineer must present his or her code and explain the risk involved.

The SPI II says the top risk is dealing with a remote sequencing engineer for the first time. It's proving harder to check each other's work if people aren't physically in the building. Noted. The process continues for every instrument.

While the science team is interested in what can go right, you might argue that the shift II engineers are looking for everything that can go *wrong*. Hanging out in these less-trod areas of Mission Control gives you more of a dismal outlook and helps explain some of the tension between scientists and engineers—even why declaring mission success now is important for some but disingenuous to others.

This issue of success comes back to the core of space exploration risk.

"When you talk about risk you have to consider that that means engineering risk. JPL limits the instruments because they see some activities as risky," Chris tells me.

"For example, Julia Bell will say we can't do a particular activity because it violates a rule, or it's a 'risk.' But what's she's saying is, it's a risk for the software. She can't point to how it risks the actual lander hardware. Because it doesn't. It's hard to find JPL people who will stand behind a risk and say 'I accept responsibility for this.' It's not worth it to them," Chris says. I ask him about Dara Sabahi, superstar engineer. He says there are lots of exceptions.

"They have some of the best engineers anywhere. But the culture makes them very risk-averse," Chris explains. That helps them land missions, but it also gets in the way when you're on the ground and trying to work. Chris says you must be precise when you talk about risk.

"When the science team wants to try new things, those aren't real risks. They're not damaging the hardware, they're just pushing the limits to get good science and explore Mars," Chris says. And that's what we want. Of course, that attitude got him disqualified as payload manager. He says JPL flatly refused to have him in the position.

The man responsible for the hardware was a JPL engineer named Mike Gross.

"He's a big, stout pit bull with a loud voice," Chris says. "We used to scream at each other about this. But we respect each other, and we're still friends." They would have the same argument about risk over and over.

"What is the risk to the hardware?" Chris says he would ask. If Mike couldn't answer, Chris would proceed.

Uplink is a place to look at everything that can possibly go wrong. When everything is looked at in terms of how it might fail, you get a very different sense of Mars exploration. It's a shock when anything works at all. After a few nights on shift II, I generally have trouble believing anything will ever work. It's just too complicated, and really depressing.

"Somedays the spacecraft is very resilient, and other days it feels so flimsy and all comes to a halt," Bob Denise says.

"I'd say it's very responsive to errors," one of the spacecraft engineers says. But in spite of all the potential problems, they have a simple plan for our second day on Earth time. We're moving closer to another TEGA delivery and opening up new territory in our workspace.

Denise signs the plan and faxes it to Denver. Everyone calls it a day. This is the process. Shift II dutifully services the scientists' every flight and fancy, putting together the plan that lets the scientists discover perchlorate, ice, water, and everything else. They don't get much glory, but there's a lot of satisfaction.

SEVERAL BOXES OF PIZZA ARRIVE AS THE PLAN WRAPS UP. IT'S WALTER Goetz's last uplink shift. He'll come in tomorrow for a bit, but then he'll head to the airport. He must get back to his teaching duties.

Katie Dunn asks if they can take a picture.

"I'm sad to see you go," she says. We all are. Walter has his arms at his sides. He's wearing a buttoned shirt, dark pants, socks, and sandals. Katie puts her arm around his waist to pose for the photo.

"If you touch me this way, my wife will get jealous," Walter says, raising his eyebrows. Katie blushes. We all laugh. Bob Denise, of all people, can't figure out how to make the flash work on the camera.

"Can one of the SSI engineers please help Bob?" Chris Swan, the strategic mission manager, asks.

AFTER SHIFT, IT'S STILL EARLY; CHRIS SHINOHARA ASKS IF ANYONE wants to grab some quick dinner and a beer. I accept.

Over pastrami sandwiches and frosty mugs of local brew, Chris speaks candidly about the problems.

"With TEGA," he says, "by the time everyone came around to the conclusion that they couldn't make it work, it was really late. The ion pump didn't work. They had to take it apart. The deeper inside you go to fix these parts, the more expensive it gets. They all have to be sealed in clean rooms. It was a mess."

The main disappointment was with how NASA handled the problems. It's the same scientist/engineer divide. Although Chris's background is in software engineering, he knows his job: he is there to serve the science team, to make the science possible. If the science doesn't get done, then your software isn't doing its job. It gets late, but Chris is trying to explain how the SSI camera works and why it's harder than you think to take great pictures on Mars.

Chris Shinohara isn't just the instrument manager for SSI who plays a tactical role as shift lead on the mission; he's also the general manager for Mission Control. That makes him pretty much responsible for everything that happens in the building. Are top-secret landing ops guarded? Ask Chris. Is there a toilet paper shortage? Ask Chris. There's a pretty wide scope for the job. When he's not managing the building, he's responsible for SSI. Chris is not a one-job type of engineer.

When he first met Peter, Chris went down to his office because he heard he was hiring. "I'd like to apply for your job postings," Chris said.

"Which one?" Peter asked.

"Both," Chris replied. Peter and Chris immediately clicked, and Peter loved his drive and enthusiasm. He got the jobs, worked on the Pathfinder mission, and they have been best friends ever since. Peter was even the best man at Chris's wedding (to their colleague Heather Enos).

The restaurant we're at wants to close; they're sweeping under our feet. But Chris ignores the broom and keeps talking. I think he can tell I'm confused about how exactly the filters on the SSI work, but he goes over everything several times. He tells me not to worry about keeping up with all the science and engineering. "It's impossible to get it all right," he says. "Just tell the story you see." That's a relief. But still I tell him I'm worried it's not flattering to NASA or that Mike Hecht might come off looking like a jerk, even though he's really a sweet guy.

"Don't worry about anyone else. Just tell your story," he says again. It's reassuring, and not just because he's a gun enthusiast and expert marksman. But he really wants people to get excited about space.

Then he goes on a rant against project executives at NASA. The same ones that everyone seems to complain about.

"They don't give a shit about the science in these missions," Chris says. "Where have they been?" he asks. I don't know, but it's true that they're rarely here. They show up when NASA is annoyed or there's some big press conference.

They probably have other projects, right?

"This is their only project," he says and motions for us to go. We get up and wander down the block. Regardless of the number of projects they have, this one should certainly be a focus. We find a bar and Chris tries to teach me more about the secret treacheries of Martian optics. It's less controversial to talk about color calibration, compression, sub-framing, and filter wheels, but a bit harder to keep up with the material.

"You need your images to be consistent regardless of conditions," Chris says, breaking it down. "Accurate color is important for the science team. That's why we needed to relentlessly protect the calibration targets from dirt. Those paint chips are a reference point and how we accurately represent color on Mars." I like to get lost in the details and pretend I understand, especially after a few beers.

Listening to Chris, I start to feel nostalgic. He and Heather had welcomed everyone to Tucson and done their best to make us feel at home while working on Mars. Chris hosted after-shift hikes, took us deep into the desert, and brought countless dinners to the SOC. Both he and Heather hosted parties to celebrate our successes and answered every question I ever asked. And tomorrow, when I'm hung over and regretting not going home and sleeping after shift, Chris will be on time and chipper. I'm being out-drunk and out-nerded by a Harley Davidson-riding, gun-toting planetary engineer. I love the people of Mars.

LATER THAT EVENING THERE'S A NEW *AVIATION WEEK* STORY POSTED online. It's from Covault.

"Phoenix Lander Mission Deemed a Success" is the title of the story. Covault scooped them! Unbelievable. The article says Phoenix declared mission success. And Covault declared it before NASA can get their press release out. I'm not even sure they decided it was a success yet. The article says: "The Phoenix Mars lander and its science and operations teams at the University of Arizona and NASA's Jet Propulsion Laboratory (JPL) will reach key milestones this week, including an official determination by NASA that the mission is a complete success . . . Designating the mission a success means it has completed key contractual and scientific goals with plenty of margin remaining in its systems for additional imaging, wet chemistry and microscope tests, and for the continued search for organics by its organic chemistry instrument."

Man, he's good.

CHAPTER THIRTY-FOUR

TWO DAYS FORWARD, ONE SOL BACK

SOL 80

I ARRIVE AT 7:45 A.M. THERE ARE ONLY A HANDFUL OF PEOPLE AT the SOC. I'm surprised it's so quiet. The transition to Earth time seemed like it would be a big relief. But it's proving difficult to readjust. Ray Arvidson warned us that it would take a month to feel normal again. Blech.

At kickoff, Ray reminds everyone we just received downlink from sol 79. We uplinked sol 80 and we're now planning sol 81. Usually we'd get the downlink from sol 80 right after kickoff. When we were on Mars time, it all made sense. Earth time messes it all up. It's some form of spaceheimer's. Today we plan and execute sol 81; but it's still sol 80 on Mars and we don't yet know what happened on 79. It's hard to know what to call it. I dread someone asking me why this chapter is called sol 80, and it's just going to get worse. Keeping track of the days gets trickier and trickier.

Today's goal is to sample from Cupboard and then deliver to TEGA. The TA-7 doors are open and ready for some dirt. If the RA hasn't safed, we're going for it. Another sample could come any day.

Mike Mellon waves hello as he walks through the SOC. Wait! Yesterday was his last day. Now time is going backwards?

"No, I couldn't stay away," he says. And mission managers needed his expertise. Now he's trying to catch up.

"I never should have left," he says. There are ten sols left in the primary mission.

"A lot of people will leave over the next week," Doug tells the team.

The spacecraft team is the first to go remote. Next week all the core spacecraft tasks like batteries, solar panels, heating, communications, and general lander wellbeing will be done at Lockheed Martin. We all applaud their effort. Ed Sedivy, who ran the operation from Colorado, and Matt Cox here at the SOC get universal praise from the engineers and scientists.

"You really get your money's worth from them. They make things happen," Chris tells me. I'm sorry I didn't spend more time with them. Yes, they've been here this whole time. An army of engineers that barely get mentioned. For shame. Twenty-four hours and forty minutes just aren't enough hours in a day.

Peter will host a special science meeting for the spacecraft engineers.

"I want them to go home with a sense of what they did for Phoenix science," Peter says. Because of a typical sol's schedule, the shift II team doesn't get to attend the science meetings. And there isn't time to follow the scientific developments. They are hyper-focused on a particular activity and don't often get to hear about what scientific discovery their engineer talents yield. Peter, ever conscious of the bigger picture, asked the theme group and instrument leads to give a recap of what we've learned so far.

I SEE CHRIS SHINOHARA IN HIS OFFICE LOOKING GLUM.

"I have to fill out layoff notices," he says. The grant that Peter and Chris were working on fell through. They hoped the State of Arizona would help fund Phoenix data analysis after the mission concluded. Instead they gave the money to a mining consortium. Space drama meets budget realities.

Filling out these notices reminds Chris how hard it was to attract talent in the early days of the mission.

"We had no money and a lot of problems," Chris says, "so we had to do some convincing to get some people on board." When the engineers heard Phoenix would use the recycled body of the Polar Lander, some balked at working on a mission that had already failed once. Somehow, they assembled a team. In the last six years, they've become a family.

"I already talked to most everyone," Chris says. "It won't be a surprise. They know we had limited funding for distributed ops."

"Still, it's not something you look forward to. But we do it and move on," he says. Even Peter says he's not sure what he'll do next. A donor endowed a chair in his name. So he's not getting kicked to the curb. He has a salary and some research money. But what will he do with it? This has been with him for so long.

"I think he's going to teach a class on Mars next semester," Chris says of Peter's plans.

Chris's office is separated from Peter's by a thick wall. While Chris and I talk about the layoffs, I can hear Peter arguing with Barry through the door. I assume Barry wants to declare mission success and, the *Aviation Week* headline not withstanding, Peter is not quite ready; but who knows.

"THE FROST IS ON THE PUMPKIN. THE LEAVES ARE CHANGING. PEOPLE are going back to school. It's time to take stock," Peter says to his team. There's a diverse audience for this special session at the end of sol science. The shift II back-room engineers who command the spacecraft are here.

Peter wants to thank the spacecraft team one more time.

"We applaud their efforts," he says. And we do.

Peter begins his Powerpoint presentation of our summer on Mars. He starts with the main premise of the mission: to land in the Arctic where Bill predicted there might be ice.

"We want to talk to the ice, but it can't talk," Peter says. "We must look for the next best thing. So we have to look at how the soils were modified by these interactions." These modifications created a mineralogical echo, a unique signature of what happened in the presence of ice. And once the ice spoke, the team needs to determine what it said about the habitability of Mars.

"We were able to show there was water ice with the TEGA measurements. And that was step one," Peter says. "Now we would need some kind of energy source. And we've found perchlorate. That result has been partially verified." There's already evidence on Earth of microbes that happily subsist on perchlorate. That qualifies as an energy source.

"We still don't know much about organics yet, but we do have nutrients," Peter says. And there are still four TEGA cells and two WCL left for testing. We followed the water and are on our way to determining the habitability of Mars. We take the next step in Mars exploration. Now there's the possibility to look for life. That should keep us coming back.

"We all know TEGA is on the spacecraft, but what does it do? Except cause trouble and send people to the penalty box," Doug Ming says. It's a great beginning to what is otherwise a highly technical talk.

"In English!" Peter says. This presentation is supposed to be for a general audience—not TEGA specialists. Somehow the scientists just can't tell us what's happening without logarithmic scales and technical terms. Peter looks annoyed.

Sam Kounaves gives the WCL roundup.

"These are very preliminary and very simplistic results, as this is a very complex system," he says. Kounaves spends a few minutes making caveats and hedging everything he is about to share. "WCL is like a tongue," he says. "It must taste soluble things. Mostly salts." Sam describes the analyses of soil samples, from the surface and one from the trench. He shows us where each sample was acquired. Then we get to the first "big" discovery in WCL—a slightly alkaline soil. Remember the asparagus days? The chemistry team found calcium, magnesium, potassium, sodium, chlorine, sulfate, and of course the big surprise, perchlorate. The team looks at how these elements combine to form salts. The types of salts that form tell us about the environment. For instance, we know that certain combinations only form in the presence of liquid water. When we piece these clues together, the results will broaden our picture of ancient Mars.

Peter looks fairly pleased. Finally, it's a presentation non-chemists can understand. Sam's second slide is a scatter plot with dense technical writing. Peter shakes his head. Logarithmic scatter plot: who can understand that?

There are more technical talks. They're too technical. And we all glean whatever bits we can, some ice cloud discovery here, more evidence of ice there. This session didn't go exactly as planned.

"On behalf of the spacecraft team, I just want to say it was an honor to work with the science team," Matt Cox says after the concluding presentation. The engineers applaud the science team. Then the science team applauds the engineers. Then everyone applauds everyone.

PETER SMITH ADMITS HE'S ACTUALLY QUITE SHY, OVER LASAGNA AT his regular spot in Tucson. With few activities this afternoon, Peter agreed to have lunch with me, and with a little less pressure from NASA and the team, there's finally hope for our bonding session.

"It took me a long time to come out of my shell," he says. "There's no point in being shy in life. You have to speak up." And then he reiterates his hope that he's not going look like some stereotypical mad scientist in this book. Somehow he sees that as a bad thing. Nerds have been cool for a long time, I say, but reassure him that I'll do my best.

"I'd come home every day and find my parents reading in their easy chairs," he says, looking up at the ceiling of this pasta joint. His arrival home from school usually coincided with cocktails and reading hour.

"It was kind of a family tradition in the Smith home. My mother and father often hosted scientific salons or impromptu mini-operas. It was not a rough-and-tumble place," he says. Peter's mother, an opera singer, would call Peter and his brother to dinner with an aria.

Peter's father was a gentle scholar. He'd spent his wild youth gallivanting around the planet curing disease. Family life was more settled and calm.

"I was the quiet kid in class who was always staring out the window, oblivious to what was happening just a few feet in front of me but lost in faraway worlds," he says. "Maybe not much has changed."

Peter's father, Hugh Hollingsworth Smith, was a bon vivant and adventurous spirit. Hugh was born in Easley, South Carolina in 1902. He left his sleepy home town to become a doctor and eventually travel the world to rid it of disease. Hugh specialized in tropical diseases and worked toward the lofty goal of eradicating the plagues of the twentieth century. He began his career as a virologist doing fieldwork in the Caribbean; then South America working as a researcher for the Rockefeller Foundation's Division for Public Health and Medicine. In 1937, while conducting field trials on attenuated strains of the yellow fever vaccine, Hugh and his boss, Max Theiler, announced their strain of yellow fever vaccine effective. Hugh and Max cured yellow fever. They saved millions of lives.

This story, a favorite of Peter's, is cause for both celebration and resentment in the Smith household. Max Theiler received the Nobel Prize for his accomplishment. Hugh Smith did not. Instead, Hugh was appointed an associate director at the Rockefeller Institute. But an office job in New York wasn't much of a consolation. He retired from the Rockefeller Institute that same year, at the ripe old age of forty-nine.

"You could say he had a dispute with management," Peter says. He married his secretary, a talented young opera singer. They packed their bags and moved to Tucson, Arizona.

Hugh knew and loved Tucson. He'd discovered it while on a road trip in the early 1920s and now was eager to start a new life there. He got a job teaching at the University of Arizona, presided over salons, and raised his two sons, Peter and his brother Robert. And he always made sure there was time to read over cocktails.

Peter Smith recalls his childhood fondly. But he's emphatic that his father belonged at the podium in Oslo with Max Theiler. We can't know what Theiler was thinking when he left out his colleague. The science journals and historical accounts of the discovery usually credit the two men working in conjunction. And in his later years, Theiler admitted that he should have shared the prize. But it was too little too late. Hugh's disappointment was an important lesson for Peter that the politics of discovery are sometimes more confusing than the discoveries themselves.

When he was in high school, Peter resolved he was going to become a great writer.

"The only problem, I wasn't all that good at it," he says. Although, to be perfectly frank, it's not true. Peter has a great gift for narrative. He knows that it is only through story that he can recast complex ideas into exciting, easily digestible bits. Without that, the science he

and so many others do is lost to the wider world. He expects his colleagues to do the same and often shakes his fists at their outrageously
confusing graphs and explanations. Before the mission Peter sat in on
all of the practice press conferences and media training sessions the
science team underwent.

"There's too much information . . . We can't understand any of that
. . . Who's going to think the study is exciting? . . . I don't even know
what that is, and I run this mission!" Peter would shout from behind
a folding table while his co-investigators practiced explaining their
instruments to an imaginary press corps.

"You're going to lose your audience in the details. You have to try
to tell a story," he would reiterate over and over again.

After high school, Peter decided to set aside his dream of being a
great writer. Science came more naturally to him. Some of the rebellious
teenage angst he felt toward his father's profession subsided. Nevertheless, the pull of counter-culture was still strong. It was the sixties.

"I was ready to leave home and start experiencing the world," Peter
says. Peter started his studies at Occidental, a small college in California. He didn't stay long.

"There was no excitement. I wanted to be part of the social revolu-
tion going on outside my window," he says. Peter transferred to the
University of California at Berkeley. There he could study physics but
also hear Allen Ginsberg and The Grateful Dead.

"I was at the first 'Be-In' in Dolores Park," Peter says with pride.
"There were people everywhere off their heads on acid." Smith strug-
gled through school. "There was so much going on, I just couldn't
compete with the more dedicated kids," he says. "They seemed to
want it more. I was happy enough to muddle through." His junior
year, everything changed.

"Before that, it was 1969 outside my window. How could I pay attention?" He couldn't. Peter was coasting along and didn't really know
what he wanted out of life. Then, midway through his sixth semester
at Berkeley, in a class on electricity and magnetism, Peter Smith found
his calling.

Peter pushes the bread off his plate and holds it up.

"Why does this plate have a reflective pattern?" Peter asks. I don't
know. "What is light? How do we experience this?" The course

material proved more exciting than the sci-fi he'd devoured as a kid. It seemed to work its strange magic and felt better than the headiness he experienced in Dolores Park.

"I wanted to understand; and the only way to do that was to get to work," he says. The professor of the electricity and magnetism class, Sumner Davis, is a teacher Peter will never forget.

"He starts class one day by saying he had never had a student in the entire history of his teaching career score so low on a first exam and then go on to get the highest grade on the midterm," Peter says. "I knew he was talking about me, because I failed the first exam and was pretty sure I got an A on that midterm," Peter says with a big grin. He got an A+.

"I went up to him after class and asked him for a job. You don't even have to pay me. I just want to work on this," Peter recalls.

"I'll pay you anyway," Professor Davis replied.

"Then he led me down to the second basement. I didn't even know there *was* a second basement," he says. They traveled down a long hallway and through a small door.

"I started to get a little nervous. He opened the door and it looked like Dr. Frankenstein's lair. There were tubes and beakers and wires," Peter tells me. "Gosh, it was a dream."

They went into a long black room bored deep into the bedrock to reduce the vibrations from street life above.

"The machine looked like a giant beast with tubes and lenses for eyes," Peter says. It was better than he could have imagined.

"Here's your desk. You start today," Professor Davis told him.

"What do I do?" Peter asked.

"You'll run the spectrometer," Davis replied.

"In my six months there, I learned more than my previous three years in university. I learned every part of that machine," Peter tells me.

PETER USED THE EXPERIENCE IN THE SECOND BASEMENT TO GET HIS next job with one of Davis's former graduate students.

"I left for Hawai'i as soon as I graduated. I rented a small boat and lived on it with a friend. We could hear the music from the bar in the harbor. It was great. And I was a spectroscopist. My mother loved to say that I was a spectroscopist," he says.

Peter loved space as a boy. He would imagine himself in the sci-fi books he read, fighting aliens on Mars. At the University of Hawai'i he lived out those boyhood fantasies.

"It was an institute for astronomy, so I decided to take some classes. I went up to the telescope on Mauna Kea, and it re-ignited my interest in planets."

"We weren't just building spectrographs there; we were building spectrographs for *space*," he says. "I learned how to build to space-hardened equipment." Equipment that goes into space needs special certification, it has to survive radiation, shaking, freezing, heating, and a host of other tests. NASA wants to see space parts with pedigree on missions it finances. You must show that your space equipment has a history of performing well in space. It can be a tricky process. He finishes his lasagna. We order dessert.

"Dana gets all fussy if I eat like this," he tells me. Dana and Peter married in the back yard of Peter's rocket-ship–shaped home in 2005. A mutual friend introduced them in 2001 just before Peter won his Mars commission. Try competing for attention with a Mars mission.

"I asked if she wanted to go SCUBA diving on our first date. Dana said she didn't know how. Neither did I," Peter says. "So we learned together. What better way to get to know someone?" They signed up for the class and had a great time together, and although it was a stretch to add SCUBA to an already overbooked Mars schedule, they somehow managed to complete the course. The final certification dive took place in the Gulf of Mexico (Arizona is not known for its scenic beaches).

"Dana called me in a panic. I was out of town for Phoenix business when the SCUBA instructor called to ask if we wanted one or two hotel rooms. She didn't know what to tell him. I said 'One!' Isn't that the point?" Peter says with a big laugh. "We've been together ever since." And very happily, his face certainly never lights up with a huge smile when I come by the SOC.

Dana is a sculptor. She spends most of her time in a studio on the edge of town where she crafts outlandish dinosaur-themed pieces. My favorite is a religious icon called "Triceratops and Child." You can see her work featured in the Tucson airport.

Dessert arrives. Peter says after living like a sailor on a boat in Hawai'i, he felt called back to Tucson. And so he returned to the University of Arizona to get his Ph.D. in optics. He did two years of his program before getting restless.

"Frankly, I got bored and started to think I didn't really need to go through the whole process," he says regretfully and takes a spoonful of tiramisu. He left with his Master's degree. In the hyper-competitive world of building space instruments, a lack of a Ph.D. was a mark against him. With so few missions, competition for commissions is highly competitive and sometimes elitist.

Peter landed a job in Tucson at the University of Arizona's Lunar and Planetary Lab in 1978. The same year, his father published his memoir, *Life's a Pleasant Institution: The Peregrinations of a Rockefeller Doctor*.

Peter returned home. He got married and had a daughter, Sara. The marriage didn't work out, but it gave him the thing he says he's most proud of.

"She's a force in the world. Truly a great heart and certainly one great thing I've done," Peter says. The Phoenix mission coincides with his thirtieth anniversary at the lab.

CHAPTER THIRTY-FIVE
PARALYZED OPS

SOL 84

NILTON RENNO RETURNS TO THE UNIVERSITY OF MICHIGAN. He sends me an email saying he will not return to the SOC, deciding it would be better to work on his "liquid water on Mars" paper in his own lab.

After the original water splashup on sol 47, Mike Hecht apologized to Nilton for his attack at the EOS.

"I retract my primary objection to your speculations," Mike said in a message, and added that he now believed the perchlorate findings made Nilton's hypothesis possible.

Then something changed his mind. Mike Hecht has new reservations and evidence that Nilton's argument is flawed. And to prove it, Hecht gave a presentation of his own at the end-of-sol science meeting. Through a series of complicated equations, Hecht presents his idea that the lander struts, where the nodules form, represent the coldest point in the area (a local minimum), and the nodules are growing because of frost, the result of a cold spot.

Mike and Nilton both agree on the presence of ice (frozen water) and vapor (gaseous water), but can't come to terms over the evidence for liquid water. The upshot is, Hecht still thinks Nilton is wrong and he should not publish his water paper.

"There isn't enough evidence," Mike Hecht says. He argues Nilton will damage the credibility of the team when he publishes his paper.

Nilton, for his part, cries foul. And one of the reasons he didn't come back is because Mike Hecht has the tools in the MECA lab to test the hypothesis for Nilton's argument about thermodynamic evidence for liquid water, and Nilton says Mike asked his team not to run the experiments.

Hecht says it's a question of resources. They don't have enough right now. With only a few weeks left in the mission, there are still a lot of chemistry results that require analysis.

None of the other team members are too eager to comment either way. They call the dispute a "distraction." Things are personal now. Nasty voicemail messages and bitter email exchanges are just the beginning.

"It's a bit awkward," one of the scientists tells me.

"Why would he do this?" Nilton asks.

Surprisingly, Nilton and Mike are good friends. Well, they *were*. They both worked at Caltech, and Mike even spent a summer living at Nilton's home, while they worked on Matador—the dust devil project.

Two years ago, a mini-rivalry developed. Nilton and Mike got into a dispute over a piece of data that Mike claimed Nilton was suppressing; it was in a calibration report one of his graduate students wrote. Nilton was insulted.

He wrote Mike Hecht to let him know.

"I told Hecht that I believed in ethics more than anything," he tells me.

Nilton replied to Hecht's emails accusing his graduate student, Manish. He included the entire list that Hecht copied on the original message. The list included Manish, his graduate student, and Miles Smith, a current Phoenix team member and postdoctoral researcher in Hecht's research group.

To Hecht, that was an unacceptable betrayal. He should have never included his students and postdocs on an email that questioned his judgment.

Since then, they are frenemies. Now this melted ice thawed their cold war. Hecht says Nilton should not be allowed to publish.

"I was careful to follow the rules, and I should be allowed to say whatever I want," Nilton says.

After all the evidence is gathered and both men rest their cases, it's Peter's decision. He'll play Judge Judy and adjudicate the dispute. But in order to do so, he needs time to understand the arguments they're making. One thing Peter doesn't have right now is a lot of free time. He still needs to keep the funding coming so Chris doesn't have to fill out more pink slips and the mission doesn't fall apart.

It's not an easy decision, either. Either Phoenix found the most interesting bit of Mars science in decades and there really is liquid water, or announcing it would be an embarrassing mistake. Some of the scientists who took up Nilton's cause have already built new Mars models based on his work. David Fisher, a Martian glaciologist on the mission, suspects Nilton's discovery is going to change how people understand the formation of the polar caps. He's already trying to model this new idea, and he has some promising results on how Nilton's discovery enables the ice caps to move over time.

"Nilton's brine might solve some of the mysterious gaps in timing," he tells me.

"It's just an idea. No one should be afraid of new ideas," Nilton says. He argues that there's no downside for Peter and the team to support him. There's certainly interesting evidence; why not explore it further? Nilton points out how he carefully followed all the "rules of the road" as the mission dictates for reporting the finding at EOS meetings, inviting any team member who wished to take part in his liquid water research as co-authors.

Hecht doesn't have a lot of time or interest in talking to me about it, so I can't really know why he's so strongly opposed. I ask him questions in the hall or in the kitchen, but I only manage a "yes" or "no" before he disappears, making me a bit biased toward Nilton (or just lazy pursuing Hecht). Especially since Nilton tried to explain his thermodynamic models and hypothesis countless times over the last few months, and with nearly infinite patience. A fair number of our conversations consisted solely of me saying "I'm not sure I understand." Eventually I made a mental image of pickle juice and antifreeze forming fantastic

Martian brines and Nilton's argument would briefly come into focus. Hecht's explanation is lots of thermodynamic equations, difficult. To Hecht's credit, he is pursuing the science in an effort to disprove Nilton's theory. And who knows what great things will come out of it. Science stories are rarely smooth.

A good science story is powerful, and Nilton works hard on his. Just like his little obelisk joke, he's interested in making sure his audience gets it. In this regard, he provides the basic axioms and they can evaluate for themselves if his conclusion is valid (or funny). Nilton has equations, but he also has a good story. This might be part of what's got Mike so upset.

"Please understand that what you are attributing to us are simple summaries of our conclusions as conveyed verbally to the media, not step-by-step analyses," Mike responded to an engineer's inquiry about not understanding his argument against liquid water. Hecht notes that no peer-reviewed journal has published Nilton's claim for liquid water on Mars.

"If Hecht is so convinced, why doesn't he publish a counterproposal?" Nilton asks.

"This is Nilton's crazy idea. The burden of proof is on him," is Hecht's response. Human nature cannot be divorced from science.

THE MISSION PLODS ON; AND SLOWLY BUT SURELY, NEW DISCOVERIES keep coming in. There were two more successful scoopfuls for MECA and TEGA. This time, TEGA detects calcium carbonate. This finding is a lot easier to digest. It's not a ground-water contaminant or rocket fuel, it's TUMS! Yes, calcium carbonate is the main ingredient in TUMS, and it's also on Mars. Perhaps an ancient alien civilization had indigestion? The scientists surmise otherwise, but they are pleased because calcium carbonate is usually formed in the presence of liquid water. This calcium carbonate did not form recently, but more likely it formed at a time when there was likely lots of water on Mars—not the tiny amounts that Nilton and Mike are arguing about. If there's lots of calcium carbonate, it could tell a story about lots of liquid water flowing on Mars. Maybe there was a giant salty ocean on Mars. This is an exciting development.

But there are new problems too. There's an issue with TEGA. A sticky valve continues to cause havoc. The gas that usually escapes through this valve is getting trapped. When the pressure builds, the signal gets all fouled up and the atom-weighing capabilities of TEGA become unreliable.

"Also, there might be an issue with the beaker temperature sensor in WCL cell three," Richard Quinn says. "It may be prudent to consider cell number two to be the last." Richard says we'll have to keep an eye on the length of the day so we can get a full run. In his opinion, they should go for their last sample in a rock-laden trench called Stone Soup.

"But that's just my opinion. Of course I'd like us to dig as deep as possible and then see what we get," Richard says.

"I spoke with Sam Kounaves and that was his opinion too," Dick Morris says. Ray Arvidson says Stone Soup is a go.

The onset of winter brings more surprises. Jim Whiteway, the Canadian weatherman who's been quietly collecting data, shows a series of photos strung together to make a movie. To everyone's amazement, it's snow falling on Mars. Whiteway and his team discovered that Mars has clouds and it snows. What? It snows on Mars. This still doesn't mean that it rains on Mars—unless Nilton has some crazy tricks up his sleeve. Snow on Mars is unexpected and an amazing discovery. We all applaud and there might even be a gasp. The seasons are changing; it's winter.

There are more and more empty desks each day. Mark Lemmon, Tom Pike, and Morten Madsen have already said good-bye.

"I'm so proud of my kids," Morten said on his last walkabout through the SOC. "They responded to every challenge we threw at them. I'm really proud."

Just last night, Christina von Holstein-Rathlou—one of Morten's students—prepared an EOS talk about frost forming on the telltale. It twinkles and shimmers when the sun reflects off the little telltale mirror—a little disco-ball first on Mars. Of course it's the signal that this is our last song of the night.

Morten was glad that his "kids" got a chance to work a Mars mission and make a real contribution. They'll be the next generation of the Danish space program. We talked about visiting each other.

The spacecraft team is long gone, managing heat, data, and everything else from their base in Denver. Equipment starts to disappear too.

"What happened to the monitors?" Clive Cook asks. "They really seem to have shrunk. I swear they looked a lot bigger yesterday."

"Yeah, you've been downgraded," Chris replied. "JPL is taking their stuff back."

When Mark Lemmon walked off from his last sol, Leslie Tamppari came over to say a collegial good-bye. She couldn't. She had to choke back her tears. She took a second to compose herself. They've worked so hard to make this happen, years of disappointment, worry, and togetherness. So many long days and then even longer sols. One struggle after another, but they did it and now it's over. Who could imagine this day four, five, or six years ago? Ice and snowstorms, calcium carbonate and maybe even liquid water! She tries again to say good-bye, but it's too much.

"Excuse me," she says. And walks off.

DOUG ARCHER IS THE SPI I. HE'S BEEN A DOCUMENTARIAN ON THE mission and now in these final moments got a chance to work a key position. He puts up a quote on the projector.

Don't cry because it's over. Smile because it happened.—Dr. Seuss

The mission is in transition. The day starts with kickmid, or is it mid-kick? No one quite knows what to call our morning meeting. Kickoff and midpoint are now one combined meeting. This is one of many Earth time changes. At kickmid, we evaluate the data downlink and then approve the day's uplink plan in one swooping science and engineering orgy.

Just a few weeks ago you couldn't get a seat, always a row of overflow standers in the back. Now there are fewer scientists, no visiting VIPs, no camera crew or journalists waiting for Sara to escort them into downlink. There's plenty of space at the big conference table. The SOC feels so empty.

Vicky Hipkin is the science lead. She shares duties with a strategic science lead to accommodate the new parallelized ops.

"This is the parallelized ops—not paralyzed. So be careful how you pronounce that," Vicky says. Parallelized means that shift I and shift II both work on a nine-to-five schedule. The shift I team is planning two sols ahead, and shift II is sequencing a plan for the next sol.

These new days start familiarly enough with a look at the historic and strategic plan, then the downlink assessment of everything Phoenix collected during its workday. Then there's a game of musical chairs. The downlink engineers swap with the sequencing engineers to solidify the next day's plan.

There's a roll call.

"SSI is remote."

"Spacecraft is remote."

"Mission manager is remote."

Vicky puts up an image of a strategic sol plan. It's already out of date.

"We're trying to work out these two sol paths. But it's proving to be very difficult," Vicky says. Parallelized ops means sequential plans must be independent of each other.

For all the talk about how the mission will be more efficient, I feel like the same activities have been on the docket for a week. We seem to languish in our Earth time transition.

"We'll get over that hump in a few days," Joel promises.

As we get closer to the lander's impending doom, I feel more stress about completing the mission and not letting our little friend go quietly into the night. We will "rage, rage against the dying of the light."

I really hope this Mars time-induced irrational sentimentality wears off soon. The RA is now unsafed and we acquired a sample called Burning Coals. It's time to deliver to TEGA. Finally. Three of the last four sols, the RA remained safe. It's not the encoder problem either. It was—gasp—human error. It turns out the RA engineers are fallible. It only took three months of working them seven days a week to determine that they're human. But we did it.

The RA downlink engineer must send out an email with something called "encoder count updates." These updates tell the sequencing engineer the initial position of the RA.

"There was an error in the email. The count in joint four was wrong. The sequencing engineer failed to double-check, and the RA safed," Joseph said with his head low at yesterday's kickmid.

It's an easy mistake and there's no formal process for this situation. Normally, the RA team has their own internal checks; a second engineer always looks it over.

"Because we're on this new schedule and there's another drive to get ice, there was no one else," Joseph explains further.

"I'd like a report on this," the mission manager says.

"But I'm not exactly sure when I'll have time," Joseph said. That's the point exactly. No time. Then we lost two days unsafing. Now we're back and the RA team recovered.

Katie Dunn is the shift lead.

"The RA is unsafed and ready to go," she announces happily.

"RA has a sample," Ashitey says. "It's about 2 to 5 cc's. I think we should proceed, but let's ask the dig czar." Ashitey suggests using the rasp to channel the dirt down the center of the scoop, since it's all collected on one side. He thinks that will aid the delivery. The good old sprinkle delivery.

"I'm a little worried about the small volume of sample," Dave Hamara says.

That elicits groans. Who can be picky at a time like this? He reconsiders.

"It's on the low end, but it should work," Hamara says. "But we would like for the RA to dump the scoop after they sprinkle to get every last grain."

They're going to go for it.

CHRIS MAKES ME A BURRITO FOR LUNCH. HE MAKES PETER ONE TOO. I ask Peter if he's up for another interview to bare his soul. It'll be very cathartic, I assure him. These are my last grasps at an emotional connection with Peter. Even though we had a promising lasagna tête-à-tête, I tell him we don't want the book to read like *Tuesday's with Nobody*.

"Oh now, we're far from that," Peter says. He walks away. With just four sols remaining, I need to dig up some Peter-themed excitement to end this book.

ASHITEY IS THE SHIFT II SEQUENCING ENGINEER. HE WORKS AT HIS machine in uplink, pausing to explain the intricacies of "guided moves" that let you over-torque versus "free-space" moves in the context of the encoder problems. This is advanced robotics.

"This encoder problem is going to hamper our efforts," Ashitey says. Each RA fault will take at least two sols to fix.

"We're trying some workarounds," he says. "But there will be more encoder faults." Even a conservative estimate of four more faults means up to twelve lost days. If there are a total of forty sols left in the mission—that's Ashitey's rough estimate—that limits the amount of science. We have to consider many of those sols will be power-limited, restricted, and have all the same issues. Soon the strategies for delivery will collide. Both MECA and TEGA want to complete their experiments, and there won't be time. There's a lot of uncertainty on what the right path will be. But if there's one thing we've learned, it's that you can't do this job without managing a bit of uncertainty.

"That's very hard for people to understand. They want everything specified. But it's not possible here," Ashitey says. He's a student of the Dara Sabahi method of space mission management.

"For all their planning on the rovers, a lot of the interesting discoveries came by accident," Ashitey says.

Katie Dunn and Byron Jones dial in to the spacecraft meeting while Ashitey and I chat. They get a busy signal and look perplexed. Who else could be calling in to a spacecraft meeting besides them!? They dial in again. The meeting starts. A scientist from Germany is giving a tour to a colleague. He opens the door, and Byron flails. The phone call is ITAR-restricted.

"You have to leave!" Byron says.

Stubbe apologizes. There's supposed to be a sign on the door when effing FNs (foreign nationals) aren't allowed in. Miles Smith picks up his head to see who he's talking about.

"Did you know there's a journalist in the room?" Miles asks. Then there's a minor panic. No one seems quite sure if I can listen in or not.

"No problem," I tell them. I'm not really a journalist anyway. But better safe than sorry. They move the call into the spacecraft room.

Before he leaves for the day, Peter agrees to meet me for one more lunch on sol 87.

CHAPTER THIRTY-SIX

THE DUDE ABIDES

SOL 87

I FEEL HUMAN AGAIN. MY BRAIN IS NO LONGER SWIMMING IN WCL calibration fluid. I eat a big breakfast with the Danes—one of our last. They head off for eye exams—part of a complete medical checkup—to be sure that the counter-fatigue study hasn't damaged them permanently with their battery of counter-fatigue tests. I go to kickmid. It's the 87th sol of the mission, but we haven't uplinked the sol 86 plan yet. It's 9:00 a.m. local Tucson time. 23:46 on Mars. So there's only fifteen minutes left of sol 85. Because we're back on Earth time, we're now two days ahead of the lander.

"How's that for confusing?" Ray says.

Kickmid gets started. Ray, the science lead, and Mark Lemmon, the strategic science lead, work remotely.

"This will be a good experiment," Ray says. Peter's assistant brings a cup of coffee to the table. Peter takes a big sip and rubs his eyes.

"Today's sol has an aggressive timeline," Ray warns. But the plan is straightforward. They have to finish by 6:15. By management fiat, the workday is limited. If the team does not finish a plan in time, they will go to a pre-made remote sensing plan.

During a pause in the action, Peter asks some of the TEGA engineers if they'll go back to their old offices on campus next week.

"I'm going to be the only one here," Peter says. "Don't go."

The core of the plan is a blind delivery TEGA and then more deep digging in the trench called Stone Soup. It's a simple plan but not much time to complete.

Uwe Keller and Diane Blaney lament the slowdown in activities. They talk with Mike Hecht about how it's only going to get worse.

"There aren't enough resources," Uwe says.

"We should still be on Mars time," Diane says.

"We're not making any progress," Uwe laments. Leslie Tamppari interrupts.

"That's not true. We've dug and safed and dug and safed and it might be slow but we're getting there," Leslie says. Mike excuses himself.

"I have to call Barry, but it shouldn't be long. He's never one to chat," Mike says.

Barry tries to help MECA get additional funding to analyze data. TEGA has similar issues getting funding. They both have more than a year of data analysis to really understand what they collected over the past ninety sols. It's not clear who is going to pay to keep the team together.

NO ONE IS SHOWING UP FOR THE END-OF-SOL SCIENCE MEETING.

"Where is everyone? Why are we even doing this?" Peter asks. The phone line is open but no one calls in.

"This is what's to come," Peter says.

"I'm here," Diane says looking up from her conversation. "We should think about the next sample."

A few more latecomers straggle in. And there's another round in the "what to sample next" discussion. They can't come to a conclusion.

"No other talks?" Peter asks when they finish.

There are none.

"Well, what about you?" Peter says. He's looking at me.

I just laugh. Good one, Peter. But then I regret not taking a turn. There's one activity I've secretly hoped for. This is my big opportunity to lobby for it. I say nothing.

There's a secret microphone on board. Well, it's not so secret but it's buried deep on the motherboard of a de-scoped instrument. I ask Bob Denise if it's possible to turn it on.

"In theory, yes," he told me. And that's good enough for me! The microphone is on an instrument called MARDI. Sadly, the Phoenix never turned it on. MARDI is a descent imager designed to take pictures of the ground as Phoenix landed. Had it worked, it would capture a series of images—the ground rising up to meet Phoenix. Why did they turn it off? MARDI shared a piece of its memory with a key landing component. There was a small chance this shared memory would cause a disaster during landing. The problem wasn't discovered until after they bolted MARDI to the spacecraft. The prudent decision was to turn it off. But there's still hope for MARDI. We could use it to listen to Mars. I want to know what Mars sounds like. Who doesn't? The science meeting ends.

CHUCK FELLOWS FROM TEGA GIVES AN UPDATE ON THE SAMPLE.

"Right now we are hopefully processing TA-7 and doing a low-temp. Tomorrow medium-temp and then high-temp. For the next sample we are developing process steps to deliver to TA-0 . . . without exposing the sample to the sun. Then we have to decide if we can get an icy sample up through the screen. Then we could choose TA-1 for an icy soil delivery," Chuck says.

"If we can't do the icy delivery, then we would open TA-6 and try to resample Rosy Red because it already has a lot of that particular sample on it," Chuck continues. That accounts for the fourth, fifth, and sixth samples.

"That leaves two more ovens," Peter says. He thinks one should be used for the organic free blank to help understand the background levels of organic material. "Then, we have one cell left. That could go to Carol's cause," he says.

"Yes, sure," Chuck says. He pauses to think. "What is Carol's cause?"

"It's a nice white sample she wants taken from Dodo-Goldilocks," Peter says.

"Didn't you see the T-shirt?" Suzanne Young asks, making light of Carol's relentless lobbying for this Dodo activity. Ever since we saw that first white chunk in Dodo, Carol insisted this was a worthwhile experiment.

THE SOL 88 PLAN IS UPLOADED WITH PLANS FOR MORE TRENCHING AND a TEGA mid-temp run. They made it another day. I ask some of the systems guys about the possibility of turning on the microphone to listen to Mars. They say it's possible and it's up to the P.I. That's Peter. I run into Peter in the hallway.

"The microphone works!" I say.

"What are you talking about? And why are we talking about this?" he asks. He's not interested in "listening" to Mars at the moment. He's more likely preoccupied with how to keep funding Phoenix research.

"Can we still have lunch?" I ask nervously.

"Yes," he responds and walks away.

WE GET INTO PETER'S CAR AND DRIVE TO A QUIET NEIGHBORHOOD.

"This little strip mall used to be the heart of downtown, back when I was a boy," Peter says. We sidle up to the bar and order a couple glasses of wine to celebrate the success of the summer.

This is it. This is our last chance to connect. The restaurant is nearly empty. It's really just us and some guy sitting at the other end of the bar.

"Great movie," I overhear the man at the bar say to the bartender.

The Big Lebowski is playing on the TV mounted above the bar. The bartender ignores the comment and continues to clean and dry an array of wine glasses.

"I'm a new student," he says. "I just moved here." The chatty guy continues to talk. The bartender continues to ignore him.

Frankly, I'm not faring much better. Our conversation is halting. Between the awkward pauses, we both look up to see what The Dude, Lebowski, and Walter are up to. I toss Peter some softball Mars questions.

"Are you happy with how things went?"

"Yes."

"Anything you wish was different?"

"No."

We watch the movie.

"So, you guys starting school?" Mr. Chatty says, leaning down the bar.

I offer a polite "No." Peter chuckles, unsure how, yet still flattered, he could be mistaken for a student.

We return to our stilted question-and-answer. We get our menus. It's a respite from the awkwardness and a means to deflect Mr. Chatty's advances.

After we order, Chatty tries again. "We are nihilists, we believe in nothing," he says, quoting the movie and laughing out loud. I ignore him but Peter takes the bait.

"I love this movie," Peter says, and in about ten seconds they're engaged in deep analysis of the *The Big Lebowski*.

And for all my efforts, three months of worry and stress and this chatty guy at the bar manages to establish an easy rapport with Peter in less than five minutes. Where did I go wrong?

All this time trying to think of the right questions, and all my efforts to sound smart about Mars, and it's all for nothing. I'm sitting between them, but no matter. They just talk through me, debating plot points.

"Did you know that Donny is supposedly just a figment of Walter's imagination?"

"Yeah, and the dude is Buddha." They carry on.

All is not lost. We actually know a lot about Peter. Every day spent on the mission is an exercise in obsessive observation. Peter is that quiet leader who we count on for safe passage through the fierce storm. He's all we could ask for to lead us to another planet.

"Why did you do this?" I ask at a pause in their conversation.

"I had the opportunity. So I wrote a proposal," Peter says.

It's a crap, stock answer. And I can't wrap my head around why Peter would have me here and then not want to tell me his story. It's more mysterious than eutectic Martian perchlorate brines.

"That's not much of an answer," I say. You don't just casually pursue a Mars mission, writing the ten million documents in your spare time just for fun.

"I want to know about the drive, the ambition that gets you through the endless list of problems and things that can go wrong, sleepless nights, missed family events, and constant battling to make it work," I say, a bit surprised.

He seems a little taken aback. Maybe it's my tone.

"Ambition? What's ambition? When you say it like that, it implies that I put my morals aside to get somewhere. I didn't do that," he says. "I worked hard for this and I always did what I believed was right."

"Maybe ambition means something much more for you. A Mars mission doesn't fall in your lap. Something made you do this. Why do it?" I ask.

"There came a point when I didn't want to be number two. In light of my father's accomplishments, I wanted make moves too," he says.

He described the moment he told his mentor, Marty Tomasko, that he was moving on. That he would build his own cameras for space. Then he interrupts himself.

"You know, I thought you were here to get my story," he says. "You should have made more of an effort." Peter tells me I'm not pushy enough and that you have to fight to get the things you want. He's right. I tell him that I came here to help him but was never certain where exactly I fit. I quit my job but had no guarantee that I could stay. He didn't make it all that easy. He says I should have said something. Woah, this is crazy, we're having a fight. Awww, our first fight.

Peter says it's my responsibility to get what I need. He's busy running a Mars mission. I say he should have had a little more faith in me.

Our fight is cathartic. Once aired, our mutual grievances kind of disappear. I tell Peter that I understand what a great risk he took to give an outsider an unfiltered look at mission life. I respect that. This approach ruffles scientists' feathers and aggressively bends NASA rules.

Still, Peter was willing to face that risk because it offers a chance to tell the Phoenix story in a different way. And no one knows if that will make things better or worse. What Peter knows, beyond a shadow of a doubt, is if NASA doesn't start telling their story in a new way, they'll turn into a dusty, dead ecosystem. Not unlike our old perception of Mars.

"I'm glad you stuck with it," Peter says. And then he tells me a story.

"I GUESS I'M INTERESTED IN EXTREMES. I WANT TO KNOW WHERE THE boundary for life is. That's an exciting place. I've been to Antarctica and it's a dry, extreme place. And then the next step for me was Mars," Peter says and then pauses to reflect on his leap.

"In the nineties when we were working on Pathfinder, I took a little break. I was on vacation for just a few days. I got back and no one was around. We were still a year away from landing Pathfinder and the IMP. And I was upstairs at the lab, sitting by myself watching a NASA press conference. They were talking about a meteorite they found in Antarctica. Then they said there was life on Mars! I'd only been gone a week. And someone found life on Mars. How the? What happened!?" Peter asks. That was the beginning of a new era. "It was a wedge of doubt. Before that, Mars was dead. You had a closed casket. And we broke open a few screws," he says. Peter thought that Pathfinder might be their chance to confirm these surprising findings. They were talking about ALH84001. A piece of Mars ejected into space and landed on Earth. There was evidence of fossilized bacteria in the meteor. That meant there was a chance for a Genesis II on Mars. It was just waiting to be discovered.

Pathfinder made great discoveries and Peter's images changed how we see Mars, but there were still no fossils. Then when Polar Lander crashed, Peter thought his career might be over. He lost every single one of his contracts.

"Starting from zero at age fifty felt so hard," he says. He didn't give up. Phoenix was his chance to get back to Mars.

"The tricky thing about the Mars story is finding the dog that didn't bark. Seeing what's not there is the key. I did think we'd see organics. It's frustrating that it's not a clear story. We don't even see contaminants. Where are the organics?" Peter asks.

Peter suspects they're burning up. They get oxidized in some kind of reaction that's not well understood. It's either a disappointment or a discovery. It depends on how you look. Time will tell what it means.

"We don't know where on Mars this thing [ALH84001] came from, but it had the same composition of material. It had carbonates. And now we see carbonates. So if that doesn't give you shivers. . . ." Peter says, raising his formidable eyebrows.

Peter isn't going to be the guy who finds life on Mars. Not on this mission. Sure, on the nights when he couldn't sleep from the stress, he might have thought about what it would be like. You have to dream big. We didn't find E.T., but we did find ice, salty brines, carbonates, perchlorate, and snow, took the first microscopic images of Mars, measured the pH of the regolith, and did the first soluble soil chemistry experiments. Not a bad step.

"I would have loved to prove them right. That there was life up there. I wanted to; but it's not that easy to do. Still, no one has proved them wrong," he says. "That keeps things interesting."

Peter helped us take another step down our long path of discovery. The random walk he thought was career—working in the lab, travels through India, building spectrometers, getting a master's in optics, pitching his own instruments—gave him the exact experience he would need to be a freelance Mars mission captain.

"You cringe when you think about some of these things we do when we're young. *Why did I go there? Why did I fly across the world with her?*" Peter asks. "And then somehow it all works out."

CHAPTER THIRTY-SEVEN
STIFF JOINTS

SOL 88

THE KITCHEN TABLE IN THE SOC IS FILLED WITH THE LEFTOVERS of a summer on Mars. There's a pair of scissors, a lunchbox, single-serve coffee packets. There's a can of organic milk. Some crackers and a few peaches. It's a curious pile. I study it for any clues into the psyche of the team members. Then I swipe a peach.

The usual science chatter in the SOC is peppered with flight departure times and scheduling final meetings. Summer is over.

Dave Hamara from TEGA comes into the kitchen.

"How's it going?" I ask.

"Not good," he says.

"End-of-school blues?" I ask.

"No. We safed," he says.

"Oh," I say.

"Yeah," he says. What happened?

"Well, I think I know," Dave explains that the valve on TEGA is still clogged and we probably just lost three days. I start to get the sense that my last days will not end with a bang; more of a robotic whimper.

I always figured the lander would die in a hail of alien gunfire or at least get incinerated by the LIDAR. Now it's kind of looking like it might just chug along slowly until the Martian winter comes and it's frozen in carbon dioxide. (Wait, that's what happened to Han Solo! And things turned out well for him. Oh, nope, he gets frozen in carbonite.)

Bob Denise says they'll try to hail the lander when it thaws in spring. But spring is almost two Earth years away.

"It's possible we make some kind of contact," he says. But even if they're able to connect after the long break, it's not likely much will work. Denise says there's a shellack covering the trace wire on the circuit boards.

"At −55c the shellack will have a phase transition. It will freeze and then when it thaws, it will be riddled with microfissures, and this will cause unpredictable failure," he explains.

But you're saying there's a chance, right? We might get lucky and find a limping but still operational Phoenix. If not, we all knew the stakes coming in: the lander was never coming back.

"Remember the good old days?" Ray Arvidson asks. "When we had a kickoff and there was a little gap to give people assignments to look at certain issues."

Well, the kickmid doesn't leave time for that, but Ray says he'd like to assign the theme leads to look at data sets for the following kickmid. There's not enough data analysis happening. Things feel a little disorganized.

Denise is now dubbed the "ice-man." He's responsible for the critical path to the second icy-soil delivery. The team is desperate for one more piece of this puzzle. If they could dig into the ice again and grab some organic material, they'd really have a beautiful new story of Mars, complete with missing organics and a strange briny water.

Ray dismisses the group.

Shoot, he forgot the weather report for the second day. It seems no one cares about the weather anymore. Thankfully, there's nothing unusual. Some dust accumulation, but it will not affect our operations.

Several scientists stop in to say good-bye. I could have sworn they'd already left. No one seems to be able to leave on their last day.

CARLA BITTER GIVES A TOUR OF THE PIT. IT'S GOING TO BE ONE OF THE last. The people who came today get to watch Joseph and Rob Bonitz test a delivery in the PIT. They watch with mouths agape. Excited kids and adult kids pressed against the metal barrier. They stand on their tippy toes or the metal bar to get closer. They point at the arm, the scoop, even Joseph and Bob, and exclaim how wonderful it all is. Real fans. I stand at the barrier with them.

The RA is arthritic and stiff. TEGA is in safe mode. Our little lander is falling apart. Phoenix M. Lander knew the stakes when she blasted off for Mars, but it's all so real. It's hard not to feel sympathy as something new seems to go wrong almost daily: a clogged valve in TEGA, a failing sensor in MECA, overheating laser in the MET. It's as if she knows the end is near.

CHAPTER THIRTY-EIGHT
SALTY LIQUID WATER TEARS

SOL 90

I GET UP EARLY, SHOWER, AND EVEN SHAVE FOR THE LAST DAY OF the primary mission. It feels important. But I know I'm only in for disappointment. Everyone is gone. The panoramic Mars images that covered the walls are gone, looted trophies from our safari. How could they have taken all these brilliant images of Mars? It's so darn sad.

"Why don't you take one?" Pat asks.

"Oh, I can have one?"

"Of course," Pat says. "You're part of the team."

This is going to look perfect in my lounge.

The conference phone line is open. There are about twenty people online. First order of business: figure out how to control the mouse on the computer that holds the plan.

Doug Ming kicks off the day's first meeting, and he has good news.

"We've had some confirmation of success that a low-temp ramp has run," Doug says. One more TEGA sample! I can leave on a happy note.

I don't want to ask any more questions. I'm tired of being explained stuff. I'm tired of wanting information from people all the time. So I just listen. I think about the lander working on Mars. What it's like up there with no one around to even see its trials and tribulations. I hang out with Joseph and Ashitey in the RA office. They are some of the lone holdouts at the SOC. They'll stay on so they can continue to test the activities in the PIT. We chat about going home.

"WHERE IS EVERYONE?" PETER ASKS.

"Gone remote," Bill responds.

"Yeah," Peter says.

It's the end-of-sol science meeting and there are only five of us. Urs Staufer is the atomic force microscope (AFM) co-investigator from the Max Planck Institute. He signed up to do a presentation of AFM sols 0-90.

"It's my pleasure to give my first AFM presentation, and sad also because it's my last day at the SOC. I want to thank everyone who worked on this," Urs says.

He shows us the first-ever nanoscale images of another planet. He's just one of the hundreds of people who did amazing things this summer. But sadly he's left out from most of this story, along with so many others who made this grand space narrative succeed. I hope someone writes several follow-ups with titles like "Urs and the AFM Monster."

It will chronicle the wacky adventures of the AFM team* as they put a sensitive piece of lab equipment on a harsh environment 200 million miles away. Amazing.

---

*Making an AFM image is a precision physical process not exactly suited for the wilds of Mars. The image is made by tracing the surface of an object with a tiny probe. A carefully calibrated tip, just a few nanometers wide and cantilevered like a record needle, is dragged over the structure you want to image. The smallest laser you can imagine measures whether or not your tip flexes as it meets new features of the particle. The pin bounces up and down as it's dragged across the surface; then somehow, the laser/flexing feedback loop is measured and adjusted and then you get an amazing 3D image of tiny things. The AFM was supposed to be a proof of concept for doing this type of imaging on Mars. Consider it proved. And there are now nano images of Mars to show for it.

Line Drube, one of my Danish roommates, gives the last end-of-science presentation I'll attend. It's a follow-up presentation on Nilton's nodule talk. She tracked the changes in the nodules since Nilton first gave his talk.

"Large areas on the lander struts now have some kinds of bumps," she says.

The two masses that Nilton first noticed grew considerably. She makes a caveat that these could be frost. It's a sensitive topic these days. Nilton is on the phone. He called in from his lab in Michigan.

"I hope that we can continue to monitor the area," Nilton says. He tells me that his paper is almost ready. His results are exciting. He's still waiting for the team's okay, but support is slowly building.

Steve Wood, one of the MECA co-investigators, suggests we knock off some bits of the ice with the TECP and collect it with the scoop.

I think that's awesome. I clap, overwhelmed by the feeling that I'm now a part of operations. Peter shoots me a look.

Stubbe, who is leaving for Germany, asks if there will be an EOS agenda so he can decide if he's going to stay up until 4 a.m.

Peter leans back in his SMITH, P.I. Captain's chair with the Star Trek font.

"Great idea," he says, firmly in command of this ship.

"Thanks for coming to the end-of-science meeting. I guess we're done here," he says.

I leave for the airport.

PHOENIX WENT TO SLEEP ON NOVEMBER 2, 2008, AND DID NOT wake up. The JPL tweeted its final remarks: 01010100 01110010 01101001 01110101 01101101 01110000 01101000. Binary code for "TRIUMPH."

CHAPTER THIRTY-NINE
SOL SEARCHING

SOL 91+

PHOENIX EVENTUALLY THAWED FROM ITS HAN SOLO-LIKE FROZEN carbon dioxide encasement. But ensuing attempts to hail the lander failed. An image of the lander, taken by the Mars Reconnaissance Orbiter, showed cracking and damage on the solar panels. Now Phoenix is Martian installation art. A little beachhead planted by humans on a faraway world that says "Peter Smith and the Phoenix team were here."

In spite of the setbacks, interruptions, complaining, conspiracies, ice cream indulgences, and general drama, the Phoenix Mars mission was a resounding success. Hundreds of science firsts and new discoveries: ice, perchlorate, nutrient-rich soil, clouds, snow, sticky soil, even liquid water. It's all detailed in the paper Peter Smith and the team published in *Science* magazine about the history of water on Mars.* Following the water turned out to be a good strategy.

*Smith, Peter. "H$_2$O at the Phoenix Landing Site." *Science.* 3 July 2009: Vol. 325, no. 5936, pp. 58-61.

The dean at the University of Arizona thought Phoenix's ground-breaking discovery—work that proved the presence of a shallow ice table, discovered all the minerals in the soil, and demonstrated evidence that water had once flowed on Mars—would make for a great thesis project. Peter's graduate school records indicated that he was only a few units shy of completing his doctoral coursework. So he took a couple classes, gathered an advisory committee, and submitted the Phoenix findings as his doctoral research. Meanwhile, Peter's daughter, Sara, pursued her degree in the university's Geography department. In 2009, Peter successfully defended his thesis. Sara did too, and the two walked across the stage together to receive their Ph.Ds. Then they took a cross-country road trip to drop Sara off at the University of North Carolina for her first teaching position. Peter returned to Tucson to teach a class about Mars and get back to work on several new space projects.

The Phoenix Mars mission ended three years ago but, of course, it was only the beginning. Pushing-the-limits-of-human-knowledge type science projects are always just the beginning. Two years after we last heard from Phoenix, I returned to Tucson to catch up with Peter. Post-Phoenix Peter is a bit more relaxed and engaging. He doesn't have NASA, JPL, and 130 other scientists to wrangle every day. Yet he's not just sitting around contemplating a memoir that corrects all the factual errors in this book. Peter is busy. He has all kinds of space projects in the works. He promises he'll tell me about them, but first it's time for small talk. He even asks about my personal life.

"Try this," Peter says. He's invented a cocktail for our meeting. It's Squirt and gin. "No, it's Diet Squirt and gin." No wonder he doesn't have much faith in my reporting. Peter lights up the grill for dinner. And I'm put in charge of chopping onions. Aww, it's just like our first meeting . . . long before Phoenix ever left the launch pad. But this time, we're shooting the breeze and someone even cries. Well, it's me and it's just the onions. But at least I can check it off the list.

"A lot has happened since sol 90. I'm glad you came back," Peter says. "You don't want to miss all that has happened since. You can't just drop us at sol 90. How would we feel?" Good point. We sit at Peter's dining room table and catch up on some science.

DR. SMITH AGREED NILTON RENNO HAD ENOUGH EVIDENCE TO PUBLISH a paper documenting Phoenix observations of liquid water on Mars.

"We only had a few images of Nilton's blobs, so it was hard to know. He did about 16 revisions of that paper. And he made a very convincing argument," Peter says. "Still, all the fighting was a bit of a sideshow. But it did lead to a lot interesting work." *The Journal of Geophysical Research* published Nilton's liquid water paper in 2009. In spite of my moral support, I was not listed as a co-author. Michael Hecht is not listed either.

"Well, he *was* included, but he and a few others asked me to take their names out," Nilton said in a followup visit. NASA headquarters invited Nilton to speak about it several times and thought his water story was a great angle for Mars.

"After a talk I gave at MIT, one of my old professors came up to me. He is a man who I really admire. This was a really tough teacher who actually failed me on my qualifying exams in graduate school. He paid me one of the greatest compliments I have ever received. He said that this observation was like what Enrico Fermi did at the test of the nuclear bomb. And then he explained that Fermi tore up a piece of paper and when he felt the blast wave, he dropped it. When he saw how far the paper fell, he computed the force of the blast because he knew the amplitude of the wave! Months later when the numbers were crunched, it turned out he was right. He had made an accurate estimate because he knew enough to make a simple test and observe," Nilton says with great pride. "He told me I'd done the same thing! And I didn't need everyone to agree with me to make my assertion. It was one of the proudest moments I've had in my career."

In spite of Peter's objection to the drama around the sideshow surrounding Nilton's findings, he's upbeat about the work.

"Nilton's blobs led to a lot of interesting findings on perchlorate. The liquid water that he saw on the legs probably had a lot to do with the effect of the force of the rockets that came at landing. The planet is not covered in liquid water. But it might be covered in perchlorate, and that might have big impact on how we measure organics," Peter says.

The perchlorate findings pointed the team in a new and interesting direction. Viking I and II searched for organics and found nothing. That led most Mars researchers to think there is no organic material and little possibility for life, at least in the equatorial regions of Mars. That was one compelling reason to send Phoenix to the nortern plains. Maybe something was different up there.The Phoenix findings might break open the possibility that Mars is a quite bit different than we imagined.

Perchlorate is a powerful oxidizer. When it's heated in TEGA (or in the thermal analyzer that Viking had) it might have destroyed any evidence of organic material. That's why Phoenix or the Viking mission didn't see it; not because it wasn't there. Chris McKay and his colleague Rafael Navarro-González decided to test this new hypothesis. They went down to the Atacama Desert to find out what would happen if they repeated the Viking experiments with soil from this desert in Chile. They added perchlorate, roughly the amount that the MECA team calculated, to the Atacama soil—which has organic and biologic material in it. When they ran the experiments, they found no traces of organics. They were gone. But they *did* see some of the same signatures from the release of chlorine. Same as Phoenix!

"That might not be proof, but it's a good fingerprint," Peter says excitedly about the future of Mars.

While the Viking experiments were redone in the hyper-arid desert of Chile, one of Peter's grad students, Doug Archer, investigated a mysterious gas release that the TEGA team saw at 300 degrees Celsius. Peter thinks it represents the moment perchlorate destroys the evidence of organics. Still more clues that Mars might indeed be covered with organic and possibly biologic material. We just can't detect it quite as easily as we originally thought.

"He thinks he can show that there's one to ten parts per million of organic material in the soil. That's like 0.0005% of the soil. Now we've got liquid water and organics," Peter says, pretending to smash his fist on the table as an exclamatory gesture. And it just gets better. Selby Cull, a student from Ray Arvidson's group, is looking into evidence that there was perchlorate all over the landing site. And maybe that will lead to showing that perchlorate is all over Mars. She thinks she sees evidence in the some of the SSI photography.

"So even if Nilton's liquid water is just a local condition, perchlorate is probably not," Peter says with enthusiasm about ongoing discovery. "This makes this mission a real stepping stone. This is a great segue for the next Mars mission, MSL [Mars Science Laboratory]. They have a way of detecting organic material by heating and another method that doesn't use heat. We don't have proof, but we're getting closer to understanding."

Perchlorate is a great story. Maybe there are microbial friends on Mars using it as a food source. And wouldn't it be an ironic twist if this very food source burned up these organics and microbes, hiding their presence from human detectors for decades. And to complicate this picture, perchlorate is extremely toxic to humans. If it's all over Mars, that would pose a big problem for human explorers of the future.

"Don't forget about the dust story. These are the highest-resolution pictures ever taken of Martian soil," Peter says, pulling together all the research into a kind of winding Mars narrative. "There's two sizes of soil. The larger particles that would feel like sand. They look like little footballs and the sand is all different colors. There are clear ones and black ones. Some people think they're volcanic glass. But the other size we see is the tiny fines; it would feel like face powder." These tiny bits in the Martian regolith would get everywhere and we would need to consider how to mitigate this dust problem for the astronauts if it has all this toxic stuff in it. Mars would be a truly dangerous and exciting place. Would this perchlorate story dampen any hopes to some day establish a colony on Mars? That question will take a long time to answer.

"The paradigm of Mars as being uninhabitable, that we have from the Viking mission, is probably wrong. The 'you can't find life on Mars' story that we hear is certainly not the full story. There's more out there, and we're on our way to finding it," Peter says. Then we eat.

"THE SCIENTISTS WERE WORRIED THAT *AVIATION LEAK*' MIGHT HAVE paid you to release data," Peter tells me when we meet the following afternoon. "So I had to keep you out of certain meetings. When we first started planning the mission, I thought it made sense to have three pillars: science, education, and promotion." The third pillar didn't go over well. There was a lot of resistance to "promoting" the mission. People were hyper-sensitive to the idea of marketing a mission to the public.

"How would you even know a mission is having a mission if they don't advertise any better than a grade 'C' movie?" Peter asks rhetorically. "No one was too excited by that idea. I thought people were going to revolt when I suggested we advertise the mission." This certainly helps explain the "awkward red-headed stepchild" of space feeling I had for the better part of the mission. The conventional wisdom is that space is so awesome, people will just flock to it without any effort to bring people in. Peter felt pressure from key members of the team every time he pursued a larger media angle for the mission. They resented the idea that he would pay someone to drive participation outside of a purely journalistic or educational play. He didn't want to have a mutiny before they even launched. So he needed to proceed with extreme caution. It's fine if they want to keep commerce out of their science, but they could at least make an effort to sell space science to their constituents.

"We wanted to do a bug's eye perspective to show what it would be like to get scooped up and baked in TEGA. Who wouldn't want to see that? But no one wanted any part of it. 'We can't say we're going to find life on Mars,' they all complained." Peter and I shake our heads.

They failed to see how this was one way to make a personal connection with the audience and let them feel like part of the discovery.

"I want to do things that people will get excited about." Peter understands in order to do that, you have to make it personal. Touch people. He clarifies that he's a scientist and not a policy man, but he thinks it's important that people are excited about how their research dollars are spent. "We need a vision. We need someone to propose something that would excite people. I think that NASA should focus on sending a manned mission to an asteroid!" Peter thinks asteroids are an exciting way to push our technical space capabilities and explore something brand-new.

"Let me show you something," Peter says as he disappears into his office. He returns and clears a space on the table. It's filled with an exam he's meant to grade this weekend. Peter shows me a movie of all the near-earth objects that have been discovered since 1980; about half a million. The visualization shows a crowded solar system and scary red blips of asteroids that cross the Earth's orbit uncomfortably often. The movie, made by a British astronomer, shows how the discoveries progress over time, most happening only in the last few years. There are no words in this movie, but the message is clear: the better we get at looking, the more we find.

"This is something people would care about because these are relevant. These are a potential threat to the Earth *and* we've never been to one. Some day one of these guys is going to be coming at us." Peter closes his laptop. "This is it. This is our exciting mission. Let's figure out how to adjust the orbit and save ourselves. You can try pushing it with thrusters, pulling it with gravity, putting a solar sail or ion engine on it. It doesn't take much to move if you have time. It would be over the course of years. Wouldn't that be a great mission? We'd be taking control of our space. We'd prove a concept. And maybe it would take two or three missions. But you'd get a lot of great research in moving these things. I think people would love it."

Peter explains how it all works. He gestures with his hands to show an asteroid being tailed by a spacecraft.

"You don't have to land on it the way you do on the moon, so it's less dangerous. You just drift over. Swing on down on some kind of tether and pull yourself back." Sounds pretty straightforward. "And of course, I'm working on some new miniature HD cameras that would film every move they'd make. Don't you think?"

That sounds amazing. One of Peter's other projects is getting miniature HD cameras space-certified so that all missions have great imagery to share with the viewers back home.

"If we're going to do something with humans, it should be interesting. The problem with the space station is that no one really knows what goes on up there. They're doing interesting things, but do you know what it is?" Peter asks me.

"Something with crystals," I say, embarrassed because I'm supposed to know these things.

"Right," Peter says. "There are interesting experiments going on there, but we can get more from our investment." Build something interesting. Peter is working on a project called Osiris Rex. It would be an unmanned trip to an asteroid. A first step toward the larger manned mission to an asteroid which Peter thinks is the right path for NASA. Osiris Rex, a well-funded Discovery-class mission. It would land on an asteroid, retrieve a sample—not just dust—and bring it back to Earth for study. A lot of the work that would go into such a mission would rocket our understanding and further planetary and deep-space missions.

"Maybe it's gonna be SpaceX. Or maybe Burt Rutan comes up with a new safe way to do things. But it's time for something new. The shuttles were designed in the late seventies. They need a warehouse of parts to keep running because no one makes them anymore," Peter tells me. "Wouldn't people like to see NASA on the cutting edge? That's part of the reason I want to build new cameras. I don't want to send 1990s cameras to Mars anymore. There's a thrill in trying to build something great for space. Maybe others are working on them and they'll get there first, but I want to build my own. I don't want to buy someone else's."

Peter will continue to work on miniature HD cameras, an asteroid mission, and even a moon-landing project over the next few years. After that, he thinks he might retire.

"Retire, but not sit around. Maybe I'll start a company or something." He objects to my insinuation that he'll just sit in a rocking chair when he moves from his rocket-ship–shaped home in the center of town out to an adobe in the hills.

AVIATION WEEK & SPACE TECHNOLOGY CLOSED ITS CAPE CANAVERAL bureau and fired many of its staffers, including Craig Covault. He now reports for an online space magazine. He maintains that his White House sources were correct.

Phoenix II was packaged up and sent to JPL for a little spruce-up before it takes its final trip to the Smithsonian in Washington. And NASA cancelled the Scout Program that gave rise to Phoenix. There won't be any more pitches for low-cost innovative freelance-led missions to Mars for the foreseeable future. The next Mars mission, Mars Science Laboratory, called "Curiosity," is scheduled to launch at the end of 2011. Stay tuned.

GLOSSARY

ACE: Individual in direct control of sending commands to the spacecraft

AFM: Atomic Force Microscope

APAM: Activity plan approval meeting

APIDs: Application process identifications

ASTG: Atmospherics Theme Group

BSTG: Biological Potential Theme Group

CAM: Command approval meeting

CSA: Canadian Space Agency

CSTG: Chemistry Theme Group

DEM: Digital elevation map

DDULT: Drop dead uplink time

EPO: Education and Public Outreach

EGA: Evolved Gas Analyzer (TEGA Component)

FCB: Faster, Cheaper, Better

GTSG: Geology Theme Group

IDE: Instrument downlink engineer

ISA: Incident Surprise Anomaly

ISAD: Icy Soil Acquisition Device

ISE: Instrument Sequencing Engineer

ITAR: International Traffic in Arms Regulations

JPL: Jet Propulsion Laboratory

LIDAR: Optical remote sensing technology (Light Detection And Ranging)

LPL:	Lunar and Planetary Laboratory (At the University of Arizona)
MECA:	Microscopy, Electrochemistry & Conductivity Analyzer
MET:	Meteorological Package
MRO:	Mars Reconnaissance Orbiter
OFB:	Organic free blank
OM:	Optical microscope
PEF:	Predicted event file
PIT:	Payload interoperability testbed
RA:	Robotic Arm
RAC:	Robotic Arm Camera
RASP:	Rapid acquisition sampling package
SOC:	Science Operations Center
SPI:	Science Plan Integrator
SSI:	Stereo service imager
TA:	Thermal analyzer (TEGA component)
TECP:	Thermal and ElectroConductivity Probe
TEGA:	Thermal and Evolved Gas Analyzer
TDL:	Tactical downlink lead
VML:	Virtual Machine Language
WCL:	Wet Chemistry Laboratory

INDEX